McGraw-Hill Education

Algebra II
Review and Workbook

McGraw-Hill Education

Algebra II
Review and Workbook

Chris Monahan

New York   Chicago   San Francisco   Athens   London   Madrid
Mexico City   Milan   New Delhi   Singapore   Sydney   Toronto

1 2 3 4 5 6 7 8 9   LHS   23 22 21 20 19 18

ISBN   978-1-260-12888-8
MHID   1-260-12888-1

e-ISBN   978-1-260-12889-5
e-MHID   1-260-12889-X

McGraw-Hill products are available at special quantity discounts to use as premiums
and sales promotions or for use in corporate training programs. To contact a
representative, please visit the Contact Us pages at www.mhprofessional.com.

To my grandsons, CR, CJ, and CT. Love you tons!

Contents

CHAPTER 1

Linear Equations and Inequalities   1

CHAPTER 2

Functions   33

Acknowledgments

I would like to thank my agent, Grace Freedson, and the editor for this project, Garret Lemoi, for their help and guidance with this project. I also need to thank my wife, Diane, for her support while I was writing this book.

Introduction

Thank you for purchasing *Algebra II Review and Workbook*. You will find that each chapter contains a large number of examples. Though each of the examples is worked out for you with descriptions of the steps as well as warnings of pitfalls to avoid, take the time to work out the problems before reading the explanations when you can. It will give you more confidence about solving problems and, should you stumble in your solution, guidance as to how the problem should be approached. I believe in the use of technology when doing mathematics, and I also believe in the power of pencil and paper. It is my firm belief that learning begins with the fingertips, travels up the arm, and works its way into the brain. Writing your thoughts on the paper requires that you give thought to all the details of the solution and often helps with the omission of steps. The calculator is there to help with the "ugly" computations needed (2 × 3 does not need a calculator, but 238.1 × 47.5 does).

An assumption has been made that you are competent using a graphing calculator. There are a few cases when the keystrokes are given (using the TI-84 and the TI-Nspire calculators). If you are unfamiliar with things like using the memory buttons on your calculator, take the time to read the manual and practice storing information on your device. (The same advice applies if you are using a graphing utility other than a Texas Instruments product.)

You will see as you go through this book and your course in Algebra II that you will develop both new skills and new concepts. Algebra II is a much different course than Algebra I in that there are many more sophisticated concepts to be learned. The most important of these concepts is that of a function. You will learn about the notation of functions, the graphs of functions, and the applications of functions.

The topics covered in this book are aligned with the Common Core State Standards for those states that adopted the program. There is more material on trigonometry than is mandated by the Common Core for Algebra II. In particular, there is a discussion of the Law of Sines, Law of Cosines, and solving trigonometric equations. This book provides comprehensive coverage of the math topics required by non–Common Core states and is also in line with the Canadian Mathematical Curricula.

Lastly, you will extend your study of probability to its application in inferential statistics. We limit the discussion at this level to basic applications of the normal distribution to make statements about the mean and proportions of populations. There is a great deal more to study about statistics in future courses.

Good luck with your studies!

CHAPTER 1

Linear Equations and Inequalities

A constant theme in the study of mathematics is to relate ideas back to concepts already learned. Linear equations and inequalities are the basic building blocks for the solution of all equations in mathematics.

Simple Linear Equations

All simple linear equations take the form $mx + n = p$ and the solution to this equation is $x = \dfrac{p-n}{m}$. As you know, the goal is to get the complicated "simple" linear equation into this basic form. The guiding principle is to gather common terms—those involving the variable in question—on one side of the equation and all other terms on the other side of the equation.

EXAMPLE

▶ Solve $4x + 5 = 3(2x - 9)$

▶ Apply the distributive property on the right side of the equation:

$$4x + 5 = 6x - 27$$

▶ Gather common terms on each side of the equation by subtracting $4x$ and adding 54 on both sides of the equation.

$$2x = 32$$

▶ Divide by 2 to get $x = 16$

Solve $\dfrac{x+5}{6} + \dfrac{3x+9}{5} = \dfrac{5x-3}{4}$

The vast majority of math teachers and students will identify the fractional expressions as the initial complication in this problem. Remove the fractions by multiplying both sides of the equation with the common denominator 60.

$$60\left(\frac{x+5}{6} + \frac{3x+9}{5}\right) = 60\left(\frac{5x-3}{4}\right)$$

Distribute the 60:

$$60\left(\frac{x+5}{6}\right) + 60\left(\frac{3x+9}{5}\right) = 60\left(\frac{5x-3}{4}\right)$$

Cancel common factors:

$$10(x+5) + 12(3x+9) = 15(5x-3)$$

Distribute:

$$10x + 50 + 36x + 108 = 75x - 45$$

Gather common terms on one side of the equation:

$$46x + 158 = 75x - 45$$

Move common terms to each side of the equation:

$$203 = 29x$$

Divide by 29:

$$x = 7$$

Check this by substituting 7 for x in the original equation:

$\dfrac{7+5}{6} + \dfrac{3(7)+9}{5} = \dfrac{5(7)-3}{4}$ becomes 2 + 6 = 8.

An equation that is written in terms of a combination of numbers and letter constants is called a literal equation. The equation $mx + n = p$ is a literal equation with x being the variable of the equation. All other letters are considered to be constants.

EXAMPLE

Solve for z: $\dfrac{z-p}{4m}+\dfrac{2z+p}{3m}=\dfrac{4z+3p}{12m}$

The variable is identified as z in the directions. Multiply both sides of the equation by the common denominator, $12m$.

$$12m\left(\frac{z-p}{4m}\right)+12m\left(\frac{2z+p}{3m}\right)=12m\left(\frac{4z+3p}{12m}\right)$$

Clear out the fractions:

$$3(z-p)+4(2z+p)=4z+3p$$

Distribute:

$$3z-3p+8z+4p=4z+3p$$

Gather common terms on one side of the equation:

$$11z+p=4z+3p$$

Move common terms to each side of the equation:

$$7z=2p$$

Divide by 7:

$$z=\frac{2p}{7}$$

There are many applications involving linear equations. Most involve system of equations and will be looked at later in this chapter. The next two examples are meant to highlight the importance of clearly defining the variable for an application and using the units of the problem to write an equation.

EXAMPLE

Ashley's dad agreed to put any dimes or quarters he received as change into a piggy bank so that she could buy a new video game. They agreed that when there were 200 coins in the bank, Ashley could have the money. They discovered that the number of quarters in the bank was 12 less than three times the number of dimes. How many coins of each type were in the bank?

"How many coins of each type were in the bank?" identifies how the variables should be labeled. If we let d represent the number of dimes in the bank, then the expression $3d-12$ represents the number of quarters. We are told there are 200 coins in the bank, so we relate the number of coins of each type in the equation $d+3d-12=200$ to answer the question. Solving this equation, we get $4d=212$ so that $d=53$. There are 53 dimes and 147 quarters in the bank.

Ashley's dad agreed to put any dimes or quarters he received as change into a piggy bank so that she could buy a new video game. They agreed that when there were 200 coins in the bank, Ashley could have the money. They discovered that when there were 200 coins in the bank, the total amount of money in the bank was $42.05. How many coins of each type were in the bank?

We can again let d represent the number of dimes in the piggy bank. All we know about the number of quarters is that they make up the remainder of the 200 coins. We identify the number of quarters as $200 - d$. The value of the money in the piggy bank is $42.05. (Rather than dealing with decimals, we state that the amount of money in the bank is 4,205 cents.) Because each dime is worth 10 cents and each quarter worth 25 cents, we can write the equation relating the value of the money: $10d + 25(200 - d) = 4{,}205$. Solve this equation:

$$10d + 5{,}000 - 25d = 4205$$

$$-15d = -795$$

$$d = 53$$

As before, there are 53 dimes and 147 quarters in the piggy bank.

EXERCISE 1.1

Solve for x.

1. $3x - 19 = 4(2x - 17)$

2. $53 - 7(2x - 3) = 11x - 49$

3. $\dfrac{x + 21}{13} + \dfrac{2x - 3}{17} = \dfrac{x - 4}{4}$

4. $\dfrac{x + b}{3a} + \dfrac{x - b}{4a} = \dfrac{2x + 3b}{5a}$

5. The garden department at The Blue Box Store is having a spring sale on plants. Stacey bought a total of 120 plants for a total cost (before tax) of $560. Stacey only bought plants that were on sale for $4 each or for $6 each. How many plants of each kind did she buy?

6. Earlene and Martin have a video library with 247 titles. They classified the movies by the era in which the movie was made. They have movies that are pre-1970s, movies from the years 1970 to 2000, and movies that were made after 2000. The number of movies they own that were made between 1970 and 2000 is 20 more than three times the number of movies they own that were made prior to 1970. The number of movies they own that were made after 2000 is 27 more than those in their library made prior to 1970. How many of each type of movie do they own?

Linear Inequalities

The most challenging piece in solving simple linear inequalities is to remember to reverse the orientation of the inequality when both sides of the sentence are multiplied or divided by a negative number.

EXAMPLE

Solve $7 - 4x > 13$. Graph the solution on a number line.

Subtract 7 to get $-4x > 6$ and divide by -4 to get $x < -1.5$. (Notice the switch in the inequality.) Because this is a strict inequality ($<$ rather than $\leq$) an open circle is used to indicate the endpoint of the set.

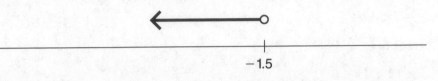

-1.5

Examine the set of numbers graphed on the accompanying number line.

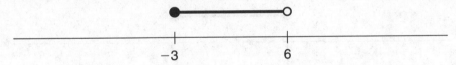

-3 6

The set contains all the points from -3, which is included in the set, through 6, which is not included. That is, using x as the variable of the inequality, $x \geq -3$ and $x < 6$. This is usually written in the more condensed form $-3 \leq x < 6$ (x is between -3, included, and 6, excluded). This is an example of a compound inequality.

There is another way of expressing intervals of numbers that require fewer symbols. Interval notation uses parentheses and brackets to denote endpoints of the intervals. The parenthesis is used to represent an open endpoint while the bracket is used to represent a closed endpoint. For example,

Inequality	Interval Notation
$-2 < x < 3$	$(-2, 3)$
$-2 \leq x \leq 3$	$[-2, 3]$
$-2 < x \leq 3$	$(-2, 3]$
$-2 \leq x < 3$	$[-2, 3)$
$x > -2$	$(-2, \infty)$
$x \geq -2$	$[-2, \infty)$
$x < 3$	$(-\infty, 3)$
$x \leq 3$	$(-\infty, 3]$

The alternate way of writing the first solution, $x < -1.5$, is $(-\infty, -1.5)$, while the solution to the second problem, $-3 \le x < 6$, is $[-3, 6)$. You probably noticed the interval for a solution open at both ends looks just like an ordered pair. You'll become comfortable with this as you take the information in context. The phrases, "plot the point $(-2, 3)$" and "the interval is $(-2, 3)$" make the usage clear.

EXAMPLE

Solve $-7 < 4x + 5 \le 13$ and graph the solution on a number line.

As seen in the above paragraph, $-7 < 4x + 5 \le 13$ means $-7 < 4x + 5$ and $4x + 5 \le 13$. The same steps will be used to solve both inequalities: subtract 5 and then divide by 4. Consequently, there is no need to split this compound inequality into its separate pieces.

$$-7 < 4x + 5 \le 13$$

Subtract 5:

$$-12 < 4x \le 8$$

Divide by 4:

$$-3 < x \le 2$$

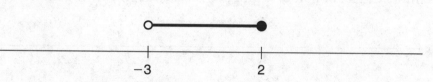

Why use interval notation here? The only good reasons are that the directions tell you to or this question is a multiple-choice question and the choices are given in interval format.

EXAMPLE

Solve $-17 < 5 - 2x \le 9$ and graph the solution on a number line.

$$-17 < 5 - 2x \le 9$$

Subtract 5:

$$-22 < -2x \le 4$$

Divide by –2:

$$11 > x \ge -2$$

Recall that when dividing an inequality by a negative number, the orientation of the inequality is reversed.

It is traditional to put the smaller number to the left when writing a compound inequality. The statement $-2 \leq x < 11$ is equivalent to the solution found above.

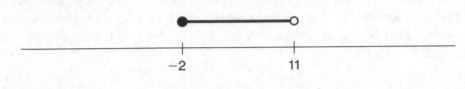

Examine the set of numbers graphed on the accompanying number line.

The set of numbers shows all those numbers that are less than -5 *or* those numbers greater than or equal to 1. Written in mathematical notation, $x < -5$ or $x \geq 1$. It is important for you to realize that the only other way to write this equality is to use interval notation and write $(-\infty, -5) \cup [1, \infty)$. (Recall the union symbol, $\cup$, is read as "or" so that any number located in one set or the other is part of the solution.) The answer $-5 < x \geq 1$ does not make any sense because -5 is not greater than 1 and $-5 < x \geq 1$ has the problem that the inequality symbols are inconsistent.

EXAMPLE

Solve $5x + 2 < 17$ or $3x - 9 > 12$. Graph the solution on a number line.

Solve each of the inequalities separately:

$5x + 2 < 17$	$3x - 9 > 12$

Subtract 2:   Add 9:

$$5x < 15 \qquad\qquad 3x > 21$$

Divide by 5:   Divide by 3:

$$x < 3 \qquad\qquad x > 7$$

The solution is written $x < 3$ or $x > 7$ in inequality notation or as $(-\infty, 3) \cup (7, \infty)$ in interval notation.

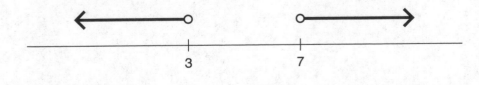

EXAMPLE

Solve $5x + 2 > 17$ or $3x - 9 < 12$. Graph the solution on a number line.

This is similar to the last problem with the change being in the direction of the inequalities. The solution to this problem is $x > 3$ or $x < 7$. The graph of this solution makes it very clear that the solution to this problem is the set of real numbers as every number on the number line is included in the solution.

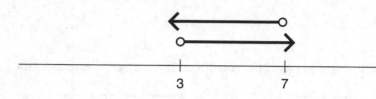

EXAMPLE

A basketball team has won 50 of the 70 games it has played. How many more games must it play and win so that the team has won at least 80% of its games?

If g is the number of extra games played and won, then $\dfrac{50 + g}{70 + g} \geq 0.8$, where

$50 + g$ represents the number of games won and $70 + g$ represents the number of games played.

Multiply both sides of the inequality by $70 + g$:

$$50 + g \geq 56 + 0.8g$$

Subtract 50 and $0.8g$ from both sides:

$$0.2g \geq 6$$

Divide by 0.2:

$$g \geq 30$$

The team must win at least its next 30 games.

EXERCISE 1.2

Solve each of the following inequalities and graph a solution on a number line.

1. $8 + 3x > 5x - 14$

2. $2(3x - 7) - 3(5 - 4x) \geq 15x + 13$

3. $17 \leq 3x - 10 < 38$

4. $-6 < 14 - 5x \leq 12$

5. $3x - 5 < 4$ or $5x - 9 \geq 16$

6. $7 - 5x < 22$ or $3x + 7 < 22$

7. A mixture of peanuts and cashews contains 30 ounces of peanuts and 20 ounces of cashews. How many ounces of cashews must be added to this mixture so that the resulting mixture is at least 60% cashews?

System of Two Linear Equations

There are a number of ways in which one can solve a system of equations. In this section, we'll look at algebraic approaches (substitution and elimination) as well as a graphical approach. Later in the chapter, we will look at a matrix approach.

ALGEBRAIC SOLUTIONS

Determining the values of the variables that make multiple equations true at the same time is important because most applications of mathematics involve the issue of meeting multiple requirements simultaneously (for example, businesspeople want to know the point at which the money they spent to put products on the market—their cost—will be gained back from the money taken in by sales—their revenue. The point at which cost = revenue is called the breakeven point.)

There are two traditional algebraic techniques for solving systems of equations—substitution and elimination. The substitution method for solving systems of equations is best applied when at least one of the equations in the system is of the $y =$ form.

EXAMPLE

Solve the system: $y = 5x + 19$
$\qquad\qquad\qquad y = 3x - 15$

Since y is equal to $5x + 19$ in the first equation, substitute this expression for y in the second equation creating the single equation in x:

$$5x + 19 = 3x - 15$$

Solve for x:

$$2x = -34$$

$$x = -17$$

▶ Find the value of y:

$$y = 5x + 19$$

$$y = 5(-17) + 19 = -66$$

▶ The solution to this system is the ordered pair $(-17, -66)$.

▶ Solve the system: $y = 2x + 25$
$\qquad\qquad\qquad 5x - 3y = 13$

▶ Substitute $2x + 25$ for y in the second equation to get:

$$5x - 3(2x + 25) = 13$$

▶ Distribute:

$$5x - 6x - 75 = 13$$

▶ Gather like terms:

$$-1x = 88$$

▶ Solve for x:

$$x = -88$$

▶ Use the $y = 2x - 5$ equation to find y:

$$y = 2(-88) + 25 = -151$$

▶ The solution to this system is the ordered pair $(-88, -151)$.

▶ If a business spends $C = 3n + 160$ dollars each day for producing n units and has revenue $R = 5n$, determine the number of units needed for the business to break even.

▶ The business breaks even when the cost equals the revenue. Therefore, set $C = R$.

$$3n + 160 = 5n$$

$$160 = 2n$$

$$n = 80$$

▶ The business must produce and sell 80 units each day in order to break even.

Although the substitution method can be used when the equations in the system are in standard form ($Ax + By = C$), the process is cumbersome and offers too many opportunities to make a mistake. The elimination (or multiplication-addition) method is a better choice. The goal in this method is to get the coefficients of one of the variables to be equal in size and opposite in sign.

EXAMPLE

Solve the system: $3x - 4y = 37$
$\qquad\qquad\quad 2x + 5y = -29$

The coefficients of the y variables are already opposite in sign. Multiplying both sides of the first equation by 5 and both sides of the second equation by 4 will get these coefficients to have equal size.

$$5(3x - 4y = 37)$$
$$4(2x + 5y = -29)$$

$$15x - 20y = 185$$
$$8x + 20y = -116$$

Add the two equations:

$$23x = 69$$

Solve for x:

$$x = 3$$

Solve for y:

$$2(3) + 5y = -29$$
$$6 + 5y = -29$$
$$5y = -35$$
$$y = -7$$

The solution to this system is the ordered pair $(3, -7)$.

EXAMPLE

Solve the system: $5x + 4y = -5$
$\qquad\qquad\quad 4x + 3y = -1$

The signs for all the coefficients are positive. Choosing to eliminate the y variable, multiply both sides of the first equation by 3 and both sides of the second equation by -4. (If you chose, you could multiply the first equation by -3 and the second by 4).

$$3(5x + 4y = -5)$$
$$-4(4x + 3y = -1)$$

$$15x + 12y = -15$$
$$-16x - 12y = 4$$

▶ Add the two equations:

$$-x = -11$$

▶ Solve for x:

$$x = 11$$

▶ Solve for y:

$$5(11) + 4y = -5$$
$$4y = -60$$
$$y = -15$$

▶ Therefore, the solution is (11, -15).

▶ Solve the system: $\dfrac{2}{3}x + \dfrac{3}{5}y = 36$

$$\dfrac{5}{6}x - \dfrac{3}{4}y = -15$$

▶ While this equation can be solved by working with the coefficients as they are, you will be less likely to make a mistake if you first rewrite the equation so that the coefficients are integers rather than fractions. To that end, multiply both sides of the first equation by 15 (the common denominator for the fractions) and both sides of the second equation by 12.

$$10x + 9y = 540$$
$$10x - 9y = -180$$

▶ Fortunately, these equations are ready to be added.

$$20x = 360$$
$$x = 18$$

▶ Solve for y:

$$10(18) + 9y = 540$$
$$9y = 360$$
$$y = 40$$

▶ Therefore, the solution is (18, 40).

▶ Please note: you should always check your answer in an original equation in case you made a mistake in an intermediate step of the problem.

EXAMPLE

Diane asked Andrew to go shopping for candy for Carson's birthday party. Andrew knows that Carson prefers Joys while some of his friends like Rounds. When he returned home, Andrew informed Diane that he bought a total of 40 candy bars and that he spent a total of $19.75. Diane asked him how much each candy bar cost, and he responded that each Joy cost $0.55 and each Round cost $0.40. After some thought, Diane said, "That should be enough of each. Thank you for going to the store." How many of each kind of candy bar did Andrew buy?

There are two types of information available in this problem: the number of candy bars purchased and the price (value) of each bar. Letting J represent the number of Joys purchased and R represent the number of Rounds purchased, the information can be displayed in two equations.

Number of candy bars:

$$J + R = 40$$

Value of the candy bars (in cents):

$$55J + 40R = 1,975$$

Solve this system of equations by substitution or elimination to determine that:

$$J = 25 \text{ and } R = 15$$

Andrew bought 25 Joys and 15 Rounds for the party.

EXERCISE 1.3

Solve each system of equations.

1. $y = 2x - 13$
   $y = -3x + 2$

2. $y = x + 7$
   $y = 4x + 22$

3. $y = x + 22$
   $2x + 3y = 16$

4. $y = x - 1.1$
   $5x + 4y = 52.3$

5. $y = x + 26$
   $7x - 2y = -113$

6. $3x - 5y = 36$
   $2x + 3y = 5$

7. $-4x + 7y = 67$
   $-3x - 2y = 14$

8. $5x - 2y = 15$
   $4x - 3y = 75$

9. $18x + 9y = -9$
   $12x + 24y = 6$

10. $3x + 4y = 42$
    $12x + 16y = 21$

11. $5x - 3y = 25$
    $15x - 9y = 75$

12. A mixture contains containing almonds and cashews is 60% almonds. If an additional 10 pounds of almonds and 20 pounds of cashews are added to the mix, the almonds will constitute 50% of the mixture. How many pounds of each type of nut were in the original mixture?

GRAPHICAL SOLUTIONS

Equations of the form $y = mx + b$ and $Ax + By = C$ are called linear equations. Since we know how to graph lines, systems of equations can be solved graphically as well as algebraically.

EXAMPLE

Solve the system $y = 3x + 7$ and $y = 5x - 9$ graphically.

Use your graphing calculator and the Intersection feature to determine the point of intersection for these lines.

When using graphing utilities, the window dimensions may need to be changed so that the point of intersection is visible on the screen.

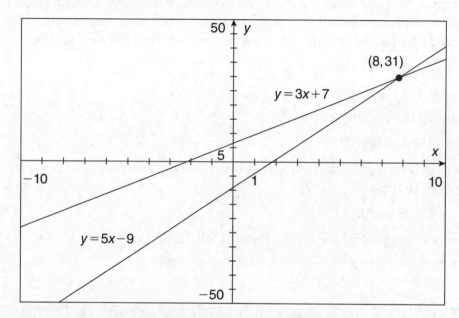

The point (8, 31) is the solution to this system of equations.

EXAMPLE

Sketch the graphs of $5x - 3y = 21$ and $y = 3x - 11$ on the same set of axes.

What are the coordinates of the point at which the graphs intersect?
When using a graphing utility, equations need to be written in the form $y =$.
Rewriting the equation, $5x - 3y = 21$ becomes $5x - 21 = 3y$ and then $y = (5x - 21)/3$ in your equation editor.

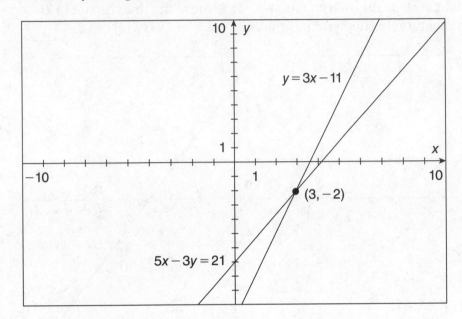

The point of intersection is $(3, -2)$.

EXAMPLE

Solve the system $4x + 2y = -3$ and $y = -2x + 8$ graphically.

When written in the form $y =$ for entry into the equation editor of your graphing utility, $4x + 2y = -3$ becomes $y = (-4x - 3)/2$. When viewed on the screen of your graphing utility, you see that the two graphs will not intersect because they are parallel. Look back at the equations. Do you see that the slope for each equation is -2? Because the lines do not intersect, the solution to this system is the empty set, written as {} or $\varnothing$.

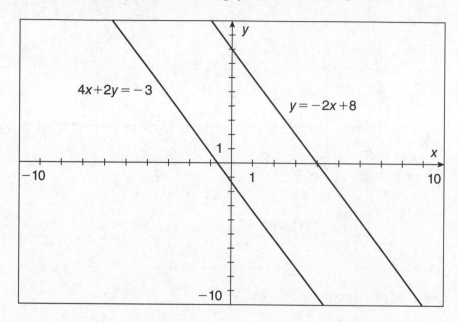

EXERCISE 1.4

Solve each of the following systems of equations graphically.

1. $y = 2x - 1$
$3x - 2y = -1$

2. $y = x - 7$
$2x - 5y = 47$

3. $5x - 2y = -16$
$3x - y = -7$

4. $3x + 4y = 56$
$-4x + 3y = -33$

5. $4x + 10y = 1$
$8x + 20y = 2$

System of Three Linear Equations

Solving a system of three (or more) linear equations algebraically requires that you pair off the equations and eliminate a variable until you get to the familiar two equations in two variables. The matrix approach to solving systems of equations speeds up the process and eliminates a lot of the tedium.

ALGEBRAIC SOLUTIONS

Equations in three variables can be graphed in a three-dimensional system—not something most classrooms have at their disposal. Equations in more than three variables do not have a physical representation available, but they do represent the ability of the users of mathematics to think in abstract terms.

In this section, you will learn to solve systems of three linear equations in three variables using the elimination method. This method can be extended to any number of equations (having the same number of variables) to find a solution (if a solution exists). The basic process is to take one of the equations and pair it against the remaining equations. The same variable will be eliminated from each of these pairs creating a new system of equations with one less equation and one less variable.

EXAMPLE

▶ Solve the system: $5x + 3y - 2z = -16$
$$2x - 4y + 3z = 41$$
$$6x + 5y + 8z = 48$$

▶ Pair off the first equation with each of the other two:

$5x + 3y - 2z = -16$ $\qquad$ $5x + 3y - 2z = -16$
$2x - 4y + 3z = 41$ $\qquad$ $6x + 5y + 8z = 48$

▶ Remove z from both systems of equations:

$3(5x + 3y - 2z = -16)$ $\qquad$ $4(5x + 3y - 2z = -16)$
$2(2x - 4y + 3z = 41)$ $\qquad$ $1(6x + 5y + 8z = 48)$

$15x + 9y - 6z = -48$ $\qquad$ $20x + 12y - 8z = -64$
$4x - 8y + 6z = 82$ $\qquad$ $6x + 5y + 8z = 48$

▶ Add:

$19x + y = 34$ $\qquad\qquad$ $26x + 17y = -16$

▶ Solve this new system of equations:

$17(19x + y = 34)$
$-1(26x + 17y = -16)$

$323x + 17y = 578$
$-26x - 17y = 16$

▶ Add:

$$297x = 594$$
$$x = 2$$

▶ Find y:

$$19(2) + y = 34$$
$$y = -4$$

▶ Find z:

$$5(2) + 3(-4) - 2z = -16$$
$$-2 - 2z = -16$$
$$z = 7$$

▶ The solution to this system of equations is the ordered triple $(2, -4, 7)$.

EXAMPLE

▶ Solve the system: $24x - 30y + 36z = -49$
$$16x + 3y - 6z = 23$$
$$8x + 12y - 6z = 28$$

▶ Pair off the third equation with each of the others.

$$24x - 30y + 36z = -49 \qquad\qquad 16x + 3y - 6z = 23$$
$$8x + 12y - 6z = 28 \qquad\qquad 8x + 12y - 6z = 28$$

▶ Eliminate z from each of the systems

$$24x - 30y + 36z = -49 \qquad\qquad 16x + 3y - 6z = 23$$
$$6(8x + 12y - 6z = 28) \qquad\qquad -1(8x + 12y - 6z = 28)$$

$$24x - 30y + 36z = -49 \qquad\qquad 16x + 3y - 6z = 23$$
$$48x + 72y - 36z = 168 \qquad\qquad -8x - 12y + 6z = -28$$

▶ Add:

$$72x + 42y = 119 \qquad\qquad\qquad 8x - 9y = -5$$

▶ Solve this system:

$$72x + 42y = 119$$
$$-9(8x - 9y = -5)$$

$$72x + 42y = 119$$
$$-72x + 81y = 45$$

$$123y = 164$$
$$y = \frac{4}{3}$$

▶ Solve for x:

$$8x - 9\left(\frac{4}{3}\right) = -5$$

$$8x = 7$$

$$x = \frac{7}{8}$$

▶ Solve for z:

$$8\left(\frac{7}{8}\right) + 12\left(\frac{4}{3}\right) - 6z = 28$$

$$-6z = 5$$

$$z = \frac{-5}{6}$$

▶ The solution to this system of equations is the ordered triple $\left(\frac{7}{8}, \frac{4}{3}, \frac{-5}{6}\right)$.

EXAMPLE

▶ Russ and Kate decided that they would put any loose change they accumulated during the day into a jar and see how much money was in the jar at the end of the month. When the month was over, Russ told Kate, "This is very interesting. We only put in nickels, dimes, and quarters. There was a total of $61 in the jar from the 398 coins we put in." "Interesting," replied Kate. Russ went on, "Isn't it? It is also true that the amount of money in the quarters we put in was three times the amount of money combined in dimes and nickels." "That means we put in . . . ," started Kate, but was interrupted by a telephone call. How many of each kind of coin did they put into the jar?

▶ There are two statements about the value of coins: the total value of coins and the relationship between the value of the quarters and the combined value of the nickels and dimes. There is also a statement about the total number of the coins. Define the variables:

> n represents the number of nickels in the jar
> d represents the number of dimes in the jar
> q represents the number of quarters in the jar

▶ The statements about the number of coins can be represented by the equations:

Total number of coins: $n + d + q = 398$

Total value of the coins: $5n + 10d + 25q = 6,100$

Quarters vs nickels and dimes: $25q = 3(5n + 10d)$

▶ Rewrite the last equation to be:

$$-15n - 30d + 25q = 0.$$

▶ The system of equations is now:

$$n + d + q = 398$$
$$5n + 10d + 25q = 6{,}100$$
$$-15n - 30d + 25q = 0$$

▶ Pair the first equation with each of the remaining equations:

$$n + d + q = 398 \qquad\qquad n + d + q = 398$$
$$5n + 10d + 25q = 6{,}100 \qquad -15n - 30d + 25q = 0$$

▶ Eliminate n:

$$-5(\, n + d + q = 398) \qquad 15(\, n + d + q = 398)$$
$$5n + 10d + 25q = 6{,}100 \qquad -15n - 30d + 25q = 0$$

$$-5n - 5d - 5q = -1{,}990 \qquad 15n + 15y + 15z = 5{,}970$$
$$5n + 10d + 25q = 6{,}100 \qquad -15n - 30d + 25q = 0$$

▶ Add:

$$5d + 20q = 4{,}110 \qquad\qquad -15d + 40q = 5{,}970$$

▶ Eliminate the d:

$$3(5d + 20q = 4{,}110)$$
$$-15d + 40q = 5{,}970$$

$$15d + 60q = 12{,}330$$
$$-15d + 40q = 5{,}970$$

▶ Add:

$$100q = 18{,}300$$

▶ Solve for q:

$$q = 183$$

▶ Finish the problem to determine that there were also 125 nickels and 90 dimes in the jar.

EXERCISE 1.5

Solve each system of equations.

1. $4x - 5y + 3z = -25$
$7x + 8y - 4z = -19$
$-3x - 4y + 5z = 37$

2. $8x - 7y + 3z = 54$
$5x + 4y + 2z = 162$
$10x + 12y - 9z = 111$

3. $7x - 5y - 2z = 39$
$4x + 3y + 2z = -65$
$2x + 5y - 12z = 79$

4. $9x + 8y - 12z = -10$
$5x + 12y + 4z = 9$
$9x - 20y - 12z = -31$

5. Tickets for the fall drama production at Eastside High School are sold at three levels—student tickets purchased in advance, student tickets purchased on the day of the performance, and adult tickets (no matter when the tickets are purchased). There are three performances for the show: Friday night, Saturday night, and a Sunday matinee. The financial report shows the following results for ticket sales: Friday's show had 250 student advance tickets, 50 student tickets sold at the door, and 300 adult tickets; Saturday's show had 600 student advance tickets, 50 student tickets sold at the door, and 500 adult tickets; Sunday's show had 50 student advance tickets, 70 student tickets sold at the door, and 250 adult tickets. Ticket receipts for the three nights were Friday $3,350, Saturday $6,150, and Sunday $2,300. What was the price charged for each type of ticket?

MATRIX SOLUTION

In the traditional approach to solving the algebraic equation $ax = b$, you divide both sides of the equation by a to get the solution $x = \dfrac{b}{a}$. Because there is no operation called division in matrix algebra, the solution to the equation

$AX = B$ is $X = A^{-1}B$, where A^{-1} is the inverse of matrix A. The matrix approach works best with a calculator that has matrix capabilities. While a matrix solution can be done with a pencil and paper approach, it will usually be much more cumbersome than the approaches shown on the earlier sections of this chapter. (See Appendix A, "An Introduction to Matrices," if you are unfamiliar with matrices.)

The matrix equivalent to the system of equations,

$$4x + 5y = 133$$
$$7x - 3y = 33$$

has the coefficient matrix $A = \begin{bmatrix} 4 & 5 \\ 7 & -3 \end{bmatrix}$, variable matrix $X = \begin{bmatrix} x \\ y \end{bmatrix}$, and

matrix of constants $B = \begin{bmatrix} 133 \\ 33 \end{bmatrix}$. The solution is $X = A^{-1}B = \begin{bmatrix} 12 \\ 17 \end{bmatrix}$, or (12, 17)

when written as an ordered pair.

The matrix solution to the system of equations,

$$6x + y - 3z = 2$$
$$3x - 4y + 5z = 65$$
$$4x - 5y - 7z = 16$$

has coefficient matrix $A = \begin{bmatrix} 6 & 1 & -3 \\ 3 & -4 & 5 \\ 4 & -5 & -7 \end{bmatrix}$, variable matrix $X = \begin{bmatrix} x \\ y \\ z \end{bmatrix}$, and

matrix of constants $B = \begin{bmatrix} 2 \\ 65 \\ 16 \end{bmatrix}$. The solution is $X = \begin{bmatrix} 4 \\ -7 \\ 5 \end{bmatrix}$, or the ordered

triple (4, –7, 5).

EXAMPLE

Use matrices to solve the system of equations

$$5x + 3y - 6z = -35$$
$$10x - 7y + 8z = 786$$
$$9x + 4y - 12z = -219$$

The coefficient matrix is $A = \begin{bmatrix} 5 & 3 & -6 \\ 10 & -7 & 8 \\ 9 & 4 & -12 \end{bmatrix}$, the variable matrix is X

$= \begin{bmatrix} x \\ y \\ z \end{bmatrix}$, and the matrix of constants is $B = \begin{bmatrix} -35 \\ 786 \\ -219 \end{bmatrix}$. The solution to the

system is $X = \begin{bmatrix} 53 \\ 48 \\ 74 \end{bmatrix}$. The ordered triple is (53, 48, 74).

A big advantage to solving systems of linear equations with a matrix solution is that the same amount of work is needed to solve a system with four equations in four variables as to solve a system of equations with two equations in two variables.

EXAMPLE

Solve the system of equations

$$5w - 4x + 12y - 3z = 363$$
$$2w + 7x - 8y + 5z = -329$$
$$9w + 12x + 15y - 7z = 902$$
$$-8w - 10x + 21y + 14z = 235$$

The coefficient matrix is $A = \begin{bmatrix} 5 & -4 & 12 & -3 \\ 2 & 7 & -8 & 5 \\ 9 & 12 & 15 & -7 \\ -8 & -10 & 21 & 14 \end{bmatrix}$, the variable matrix is

$X = \begin{bmatrix} w \\ x \\ y \\ z \end{bmatrix}$, and the matrix of constants is $B = \begin{bmatrix} 363 \\ -329 \\ 902 \\ 235 \end{bmatrix}$. The solution is

$X = \begin{bmatrix} -17 \\ 23 \\ 37 \\ -32 \end{bmatrix}$, or $w = -17$, $x = 23$, $y = 37$, and $z = -32$. The coordinates for

the solution are the ordered 4-tuple $(-17, 23, 37, -32)$.

EXERCISE 1.6

Use matrices to solve each system of equations.

1. $6x - 7y = -101$
$5x + 3y = 172$

2. $6x - 3y = 39$
$4x + 7y = 539$

3. $8x + 5y + 6z = 161$
$3x - 4y - 5z = -313$
$-4x - 13y + 7z = 974$

4. $5w - 3x + 6y - 7z = 177$
$8w + 12x + 6y - 5z = 65$
$6w + 3x - 5y + 3z = -69$
$w - x + y - z = 32$

5. Tickets for the spring musical at Mountainview High School are $8 for student tickets in advance of the performance, $10 for student tickets sold on the day of the performance, $12 for adult tickets sold in advance, and $15 for adult tickets sold on the day of the performance. The financial report after the musical performances showed that 1,950 tickets were sold with receipts totaling $22,100. The number of adult tickets sold exceeded the total number of student tickets sold by 350, and the ticket sales for the adult tickets exceeded the ticket sales for student tickets by $7,900. How many tickets of each type were sold for the musical?

System of Linear Inequalities

When graphing inequalities on a number line, the difference in the graphs of $x > 1$ and $x \geq 1$ is to use an open circle at 1 for $x > 1$ (indicating that the endpoint is not included—this is called a half line) and a closed circle for $x \geq 1$ (indicating that the point is included—this is called a ray). In the number plane, the graph of $x \geq 1$ would be a vertical line with the region to the right of the line shaded. The graph of $x > 1$ would be a dotted line to show that the boundary is not included with the region to the right shaded.

EXAMPLE

▶ Sketch the common solution for the system of inequalities:

$$y \leq 3x + 5$$
$$y > 7 - 2x$$

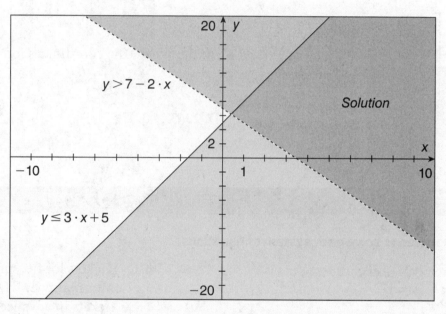

▶ The common solution is the overlapping region and is shaded a dark gray.

EXAMPLE

▶ Part of Stockwell's Gaming System business philosophy is to be sure that she always has the latest version of her favorite game Just Dance available for both Xbox and PlayStation. She always has at least one version of a game for each platform, but she never has more than a total of 20 at any one time. Demonstrate the number of possible combinations of the game that should be found on the shelves at Stockwell's store.

▶ Let x represent the number of copies of the game for Xbox and y represent the number of copies available for PlayStation.

▶ "She always has at least one version of a game for each platform" translates to the inequalities $x \geq 1$ and $y \geq 1$.

▶ "She never has more than a total of 20 at any one time" translates to $x + y \leq 20$.

▶ Graph the three inequalities on the same set of axes.

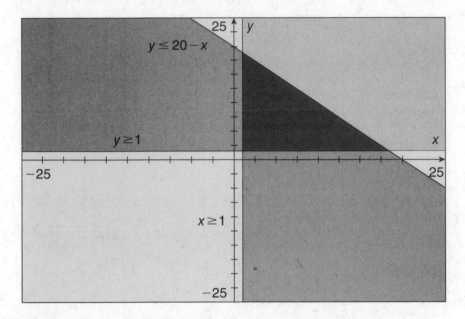

▶ The screen can get crowded when graphing more than two inequalities as shown in the image.

▶ If only the common region is shaded, the common solution looks like the next image.

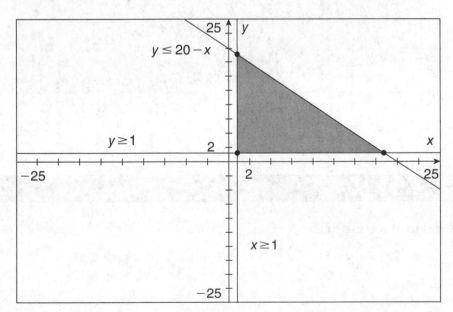

▶ The possible combinations of the number of versions of Just Dance in the Xbox or PlayStation format are represented by all the points on or inside the triangle with integer coordinates. (You consider only the integer values because, as an example, you cannot have 1.5 copies of the game available to sell to a customer.)

▶ It is good practice to shade only the common region when graphing a system of inequalities when more than two graphs are involved.

▶ Determine the common solution to the system of inequalities

$$x \geq 0 \quad y \geq 0$$
$$x + 6y \geq 18$$
$$3x + 5y \geq 30$$
$$4x + y \geq 12$$

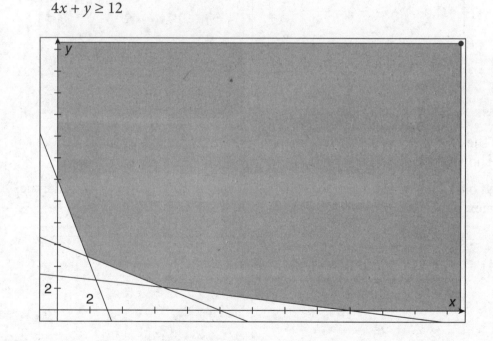

EXERCISE 1.7

Sketch the solution to each system of inequalities.

1. $x \geq 2$
    $y \leq 5$
    $y \geq x$

2. $x + y \leq 10$
    $2x - y \leq 3$
    $y \leq 7x$

3. $3x + 4y \leq 24$
    $4x + 3y \geq 12$
    $x \leq 5$

Absolute Value Equations

Most students learn the concept of absolute value as the magnitude of the number without the sign. That is, $|4| = 4$ and $|-4| = 4$. It will help you to see that both these numbers are 4 units from the origin. In fact, the **geometric definition for absolute value** is the distance a point is from the origin on the number line. This definition will prove helpful in solving absolute value equations and inequalities. An algebraic definition for absolute value is based on our knowledge of square roots: $\sqrt{x^2} = |x|$. (We'll discuss more about this when we examine quadratic functions.)

EXAMPLE

Solve $|x| = 4$

As seen above, $x = \pm 4$.

EXAMPLE

Solve $|x + 3| = 4$

In this case, $x + 3 = \pm 4$. Solve $x + 3 = -4$ to get $x = -7$ and solve $x + 3 = 4$ to get $x = 1$. Observe that -7 and 1 are each 4 units from -3 and that the solution to $x + 3 = 0$ is $x = -3$. In other words, the solution to the equation $|x + 3| = 4$ is found by sliding the solution of $|x| = 4$ to the left 3 units.

EXAMPLE

Solve $|x - 4| = 5$

Algebraic approach: $x - 4 = \pm 5$. Solve each of these equations to get $x = -1$ or $x = 9$.

Geometric approach: The solution to $x - 4 = 0$ is $x = 4$. The points 5 units from 4 on the number line are -1 and 9.

EXAMPLE

Solve $|2x - 7| = 9$

Algebraic approach: $2x - 7 = \pm 9$. Solving each of these equations, $x = -1$ or $x = 8$.

Geometric approach: The coefficient of 2 has an impact on the solution. First, factor the 2 to get $2|x - 3.5| = 9$. Divide by 2 to get $|x - 3.5| = 4.5$. The solution to $x - 3.5 = 0$ is $x = 3.5$. Those points that are 4.5 units from 3.5 on the number line are -1 and 8.

> You can use $|a - b| = |b - a|$ when solving absolute value equations or inequalities.

Solve $|12 - 4x| = 16$

Algebraic approach: $|12 - 4x| = |4x - 12|$. The problem can be written as $4x - 12 = \pm 16$. Solving each of these equations, $x = -1$ or 7.

Geometric approach: Factor -4 from inside the left side of the equation to get: $|-4(x - 3)| = 16$

Separate the factors:

$$|-4|\,|x - 3| = 16$$

Divide by $|-4|$:

$$|x - 3| = 4$$

Those points that are 4 units from 3 are -1 and 7.

EXERCISE 1.8

Solve each of the following absolute value equations.

1. $|x + 5| = 2$

2. $|x - 15| = 7$

3. $|2x - 3| = 7$

4. $|5x - 3| = 7$

5. $|11 - 4x| = 21$

6. $|31 - 2x| = 9$

Absolute Value Inequalities

If the solution to the equation $|x| = a$ are those points a units from 0 on the number line, then it stands to reason that the solution to $|x| > a$ are those points more than a units from 0 while the solution to $|x| < a$ are those points less than a units from 0. You'll find it very helpful to think of inequalities with absolute values geometrically rather than trying to memorize a set of algebraic rules for the process.

Solve $|x| \geq 6$

We are trying to find the set of points that are at least (because of the greater than *or* equal to) 6 units from 0. These points are $x \leq -6$ or $x \geq 6$.

EXAMPLE

Solve $|x - 5| \geq 6$

We are now trying to find the set of points that are at least 6 units from x = 5 (because the solution to $x - 5 = 0$ is $x = 5$). 6 units to the left of 5 is –1 and 6 units to the right of 5 is 11. Therefore, the solution to the inequality is $x \leq -1$ or $x \geq 11$.

EXAMPLE

Write an inequality with absolute value to represent the set of numbers shown on the number line.

Midway between –3 and 5 is 1. (The average of –3 and 5 is 1.) Both –3 and 5 are 4 units from 1, and the remaining points shown are more than 4 units from 1. Therefore, the absolute inequality that describes this set of numbers is $|x - 1| \geq 4$.

EXAMPLE

Solve $|4x + 12| < 10$

We know how to interpret statements of the form $|x - c| < a$: the set of points less than a units from c. Consequently, we need to factor the coefficient of x from the left to get back to the familiar form of the problem.

Factor the 4:

$|4||x + 3| < 10$

Divide by $|4|$:

$|x + 3| < 2.5$

The set of points less than 2.5 units from $x = -3$ is $-5.5 < x < -0.5$.

EXAMPLE

Solve $|3x + 14| \leq 8$

Factor the 3:

$$\left| 3\left(x + \frac{14}{3} \right) \right| \leq 8$$

Divide:

$$\left| x + \frac{14}{3} \right| \leq \frac{8}{3}$$

The solutions to the problem are those values that are at most $\frac{8}{3}$ units from $\frac{-14}{3}$:

$$\frac{-14}{3} - \frac{8}{3} = \frac{-22}{3} \text{ and}$$

$$\frac{-14}{3} + \frac{8}{3} = \frac{-6}{3} = -2.$$

Therefore, the solution to the problem is $\frac{-22}{3} \leq x \leq -2$.

EXAMPLE

An invitation says that the party will begin at 3 p.m. Social convention states that in order to arrive "on time," one should arrive within 10 minutes of the stated starting time. According to this convention, what is the acceptable interval of time during which one can arrive on time?

Within 10 minutes of the stated time allows the guest to arrive anywhere between 2:50 and 3:10 p.m.

EXAMPLE

The quality control department works under the guidelines that a process is working properly if the specs for the product are within 3 standard deviations of the mean. Suppose the mean diameter of golf ball is 3.81 cm with a standard deviation of 0.01. Write an absolute inequality for the range of diameters of the golf balls produced that the quality control process will claim are acceptable.

The statement claims so long as the diameters are within 3 times .01 cm of 3.81 cm, everything is fine. This is equivalent to the absolute value inequality $|d - 3.81| \leq 3(.01)$.

EXERCISE 1.9

Solve each of the following inequalities.

1. $|x - 3| \leq 7$ **2.** $|x + 3| > 2$ **3.** $|4x - 3| \geq 7$ **4.** $|7 - 3x| < 4$

Write an absolute value inequality that describes each of the sets graphed below.

5.

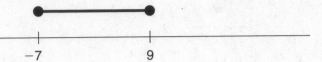

6.

CHAPTER 2

Functions

The language of mathematics is fairly specific. That is to say, it is more often the case that one cannot use synonyms when using mathematical terminology. It is because of this specificity that mathematics is considered to be a universal language. The terminology and symbols used do not lend themselves to misinterpretation.

A crucial concept is that of relations and functions.

Relations and Inverses

A relation is any set of ordered pairs. The set of all first elements (the input values) is called the *domain*, while the set of second elements (the output values) is called the *range*. Relations are traditionally named with a capital letter. For example, given the relation:

$$A = \{(2, 3), (-1, 5), (4, -3), (2, 0), (-9, 1)\}$$

The domain of A (written D_A) is $\{-9, -1, 2, 4\}$. (The domain was written in increasing order for the convenience of reading, but this is not required.) The element 2, which appears as the input for two different ordered pairs, needs to be written just the one time in the domain. The range of A (written R_A) is $\{-3, 0, 1, 3, 5\}$.

The inverse of a relation is found by interchanging the input and output values. For example, the inverse of A (written A^{-1}) is

$$A^{-1} = \{(3, 2), (5, -1), (-3, 4), (0, 2), (1, -9)\}$$

Do you see that the domain of the inverse of A is the same set as the range of A and that the range of the inverse of A is the same as the domain of A? This is very important.

EXERCISE 2.1

Given the relationships:

$A = \{(-1, 7), (6, 3), (4, 0), (4, 10), (3, 12)\}$

$B = \{(10, -1), (0, -2), (7, -10), (3, 5), (7, -2), (5, 1)\}$

$C = \{(\text{Kristen}, 5), (\text{Stacey}, 21), (\text{Kate}, 9), (\text{Colin}, 8), (\text{Carson}, 12), (\text{Brendon}, 15), (\text{Russ}, 12), (\text{Andrew}, 17)\}$

1. Find the domain of A.

2. Find the range of A.

3. Find the domain of B.

4. Find the range of B.

5. Find the domain of C.

6. Find the range of C.

7. Find A^{-1}.

8. Find B^{-1}.

9. Find C^{-1}.

Functions

All functions are relations, but not all relations are functions.

Functions are a special case of a relation. By definition, a function is a relation in which each element of the domain (the input values) has a unique element in the range (the output value). In other words, for each input value there can be only one output value. Looking at the three relations above, you can see that A is not a function because the input value of 4 is associated with the output values 0 and 10. The relation A^{-1} is a function because each input value is paired with a unique output value. (Don't be confused that the number 4 is used as an output value for two different input values. The definition of a function does not place any stipulations on the output values.)

EXAMPLE

Given the relations:

$B = \{(3, -7), (0, 2), (9, -10), (3, 5), (6, -2), (5, -1)\}$

$C = \{(\text{Kristen}, 5), (\text{Stacey}, 21), (\text{Kate}, 9), (\text{Colin}, 8), (\text{Carson}, 12), (\text{Brendon}, 15), (\text{Russ}, 12), (\text{Andrew}, 17)\}$

(a) Determine if the relation represents a function.

(b) Determine if the inverse of the relation is a function.

Solution:

$B = \{(3, -7), (0, 2), (9, -10), (3, 5), (6, -2), (5, -1)\}$

B is not a function because the input value 3 has two output values, –7 and 5.

$$B^{-1} = \{(-7, 3), (2, 0), (-10, 9), (5, 3), (-2, 6) (-1, 5)\}$$

▶ B^{-1} is a function because no input has multiple output values.

$$C = \{(\text{Kristen, 5}), (\text{Stacey, 21}), (\text{Kate, 9}), (\text{Colin, 8}), (\text{Carson, 12}),$$
$$(\text{Brendon, 15}), (\text{Russ, 12}), (\text{Andrew,17})\}$$

▶ C is a function because each input value has a unique output value.

$$C^{-1} = \{(5, \text{Kristen}), (4, \text{Stacey}), (9, \text{Kate}), (8, \text{Colin}), (12, \text{Carson}), (15,$$
$$\text{Brendon}), (12, \text{Russ}), (17, \text{Andrew})\}$$

▶ C^{-1} is not a function because the input value 12 has two output values, Carson and Russ.

EXERCISE 2.2

Given the relationships:

$$A = \{(3, 7), (-5, 9), (-5, 2), (11, 9), (13, 2)\}$$
$$B = \{(8, 7), (12, 10), (-19, 21), (12, 11), (16, 2), (25, -7)\}$$
$$C = \{(1, 1), (2, 3), (6, -1), (5, -3), (4, 0)\}$$

1. Which of the relations A, B, and C are functions?

2. Which of the relations A^{-1}, B^{-1}, and C^{-1} are functions?

3. A relation is defined by the sets {{students in your homeroom}, {e-mail addresses at which they can be reached}}. That is, the input is the set of students in your homeroom and the output is the set of e-mail addresses at which they can be reached. Must this relationship be a function? Explain.

4. Is the inverse of the relation in problem 3 a function? Explain.

5. A relation is defined by the sets {{students in your homeroom}, {the student's Social Security number}}. Must this relationship be a function? Explain.

6. Is the inverse of the relation in problem 5 a function? Explain.

7. A relation is defined by the sets {{students in your homeroom}, {biological mother}}. Must this relationship be a function? Explain.

8. Is the inverse of the relation in problem 7 a function? Explain.

Function Notation

Function notation is a very efficient way to represent multiple functions simultaneously while also indicating domain variables. Consider the function $f(x) = 9x - 2$. This is reads as "f of x equals $9x - 2$." The name of this function is f, the independent variable (the input variable) is x, and the output values are computed based on the rule $9x - 2$. In the past, you would have most likely just written $y = 9x - 2$ and thought nothing of it. Given that, be patient as you work through this section. What is the value of the output of f when the input is 4? In function notation, this would be written as $f(4) = 9(4) - 2 = 34$. Do you see that the x in the name of the function is replaced with a 4—the desired input value—and that the x in the rule of this function is also replaced with a 4? The point (4, 34) is a point on the graph of this function. Consequently, you should think of the phrase $y =$ whenever you read $f(x)$. That is, if the function reads $f(x) = 9x - 2$ you should think $y = f(x) = 9x - 2$ so that you will associate the output of the function with the y-coordinate on the graph. $f(-2) = 9(-2) - 2 = -20$ indicates that when -2 is the input, -20 is the output and the ordered pair $(-2, -20)$ is a point on the graph of this function.

> In essence, function notation is a substitution-guided process. Whatever you substitute within the parentheses on the left side of the equation is also substituted for the variable on the right-hand side of the equation.

EXAMPLE

Given $k(n) = -3n^2 + 12n + 8$. Find:

(a) $k(-3)$

(b) $k(2t - 3)$

Solutions:

(a) $k(-3) = -3(-3)^2 + 12(-3) + 8 = -27 - 36 + 8 = -55$

(b) Since $2t - 3$ is inside the parentheses, you are being told to substitute $2t - 3$ for n on the right-hand side of the equation.

$$k(2t - 3) = -3(2t - 3)^2 + 12(2t - 3) + 8$$
$$= -3(4t^2 - 12t + 9) + 24t - 36 + 8$$
$$= -12t^2 + 36t - 27 + 24t - 36 + 8$$
$$= -12t^2 + 60t - 55$$

EXERCISE 2.3

Given $f(x) = -7x + 12$, find:

1. $f(4)$

2. $f(n + 4)$

Given $g(x) = \dfrac{4x + 3}{x - 2}$, find:

3. $g(7)$

4. $g(t + 2)$

Given $p(t) = \sqrt{8t + 9}$, find:

5. $p(5)$

6. $p(r - 3)$

Arithmetic of Functions

Arithmetic can be performed on functions. For example, let $g(x) = 10x + 3$ and $p(x) = \dfrac{5x-3}{x-2}$. To calculate $g(3) + p(3)$, you first evaluate each of the functions, $g(3) = 33$ and $p(3) = 12$, and then add the results: $g(3) + p(3) = 22 + 12 = 45$. The expression $g(4) - p(1)$ shows that the input values do not have to be the same to do arithmetic: $g(4) = 43$ and $p(1) = -2$, so $g(3) - p(1) = 43 - (-2) = 45$.

What does $p(g(3))$ equal? A better question to answer first is, what does $p(g(3))$ mean? Since $g(3)$ is inside the parentheses for the function p, you are being told to make that substitution for x in the rule for p. It will be more efficient (and less writing) if you first determine that $g(3) = 33$ and evaluate $p(33)$. $p(33) = \dfrac{5(33)-3}{33-2} = \dfrac{162}{31}$. Therefore, $p(g(3)) = \dfrac{162}{31}$. Evaluating a function with another function is called **composition of functions**. While $p(g(3)) = \dfrac{162}{31}$, $g(p(3)) = g(12) = 10(12) + 3 = 123$. This illustrates that you must evaluate a composition from the inside to the outside.

> The composition $p(g(x))$ can also be written as $p \circ g(x)$.

Given $f(x) = 5x^2 - 3$ and $g(x) = \dfrac{4x-1}{2x+3}$, evaluate:

(a) $f(4) + g(-1)$

(b) $g(-2) - f(3)$

(c) $f(g(-1))$

(d) $g(f(0))$

Solutions:

(a) $f(4) = 5(4)^2 - 3 = 77$ and $g(-1) = \dfrac{4(-1)-1}{2(-1)+3} = \dfrac{-5}{1} = -5$ so $f(4) + g(-1) = 77 - 5 = 72$

(b) $g(-2) = \dfrac{4(-2)-1}{2(-2)+3} = \dfrac{-9}{-1} = 9$ and $f(3) = 5(3)^2 - 3 = 45 - 3 = 42$ so $g(-1) - f(-1) = 9 - 42 = -33$.

(c) $f(g(-1)) = f(-5) = 5(-5)^2 - 3 = 122$.

(d) $g(f(0)) = g(-3) = \dfrac{4(-3)-1}{2(-3)+3} = \dfrac{-13}{-3} = \dfrac{13}{3}$.

As you know, there are two computational areas that will not result in a real number answer: (1) do not divide by zero, and, (2) do not take the square root (or an even root) of a negative number. This is useful when trying to determine the domains of functions.

EXAMPLE

Find the domain for each function.

(a) $f(x) = \dfrac{8x + 16}{2x - 3}$

(b) $q(x) = \sqrt{12 - 3x}$

Solutions:

(a) To avoid dividing by zero, $2x - 3 \neq 0$ so $x \neq 1.5$

(b) $12 - 3x$ cannot be negative, so it must be the case that $12 - 3x \geq 0$ so that $4 \geq x$. Another way to write this is that $x \leq 4$.

Finding the range of a function is more challenging. This topic will be brought up throughout this book as particular types of functions are studied.

A key concept in economics is the notion of the *breakeven point*. It costs money to produce the goods that are going to be sold. The revenue (income) earned from selling the items produced is typically calculated by multiplying the price per unit by the number of units sold. When the seller makes back all the money spent in the production process (that is, Cost = Revenue), the seller is said to break even. The difference between revenue and cost is profit ($P = R - C$).

EXAMPLE

A manufacturer determines that her daily cost function is $C(n) = 1.5n + 1{,}250$, where n is the number of units produced and C is the number of dollars spent. If her revenue function is $R(n) = 14n$, determine her breakeven point.

To determine the breakeven point, set $R = C$. $14n = 1.5n + 1{,}250$ so that $n = 100$. Her breakeven point requires that she sell 100 units each day to retrieve the $1,400 spent on production.

EXERCISE 2.4

Use these functions to answer questions 1–6.

$$f(x) = 2x^2 - 3x + 5$$

$$g(x) = \sqrt{5x + 10}$$

1. $f(-2) + g(3)$

2. $g(f(2))$

3. $f(3)*g(0)$

4. $\dfrac{f(10)}{g(43)}$

5. $f(g(123))$

6. $g(33) - f(-2)$

Find the domain for each of the following functions.

7. $g(x) = \dfrac{5x + 1}{3x - 4}$

8. $k(x) = \dfrac{2x - 3}{x^2 - 4}$

9. $b(x) = \sqrt{4x + 36}$

10. $a(x) = \sqrt{\dfrac{-2}{3}x + 6}$

11. A manufacturer determines that his daily cost is given by $C(n) = 150n + 25{,}000$. If he sells each unit for \$400, what is the manufacturer's breakeven point?

Transformation of Functions

There are numerous ways that functions can be altered. You can slide them, flip them, stretch them, and twist them. At this level of study, we'll just consider some of the basic transformations. We'll slide them right, left, up, or down. We'll flip them over an axis (or a line parallel to an axis), and stretch them from an axis.

Look at the following graph of $p(x) = x^2$.

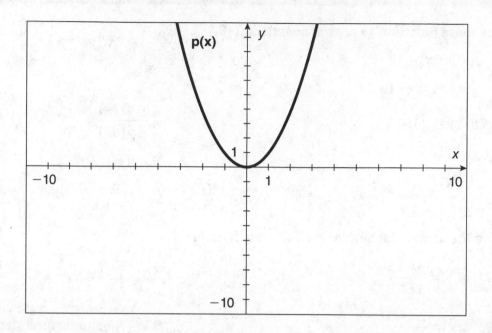

The functional notation involving transformations looks confusing—or at least, contradictory—at first. We claim that the strength of mathematical notation is to make concepts clearer, so let's take a moment to show that the notation is consistent.

Let $g(x) = p(x - 3)$, $k(x) = p(x + 3)$, $m(x) = p(x) - 3$, and $n(x) = p(x) + 3$. The graph of $p(x)$ is translated to the right 3 units to get the graph of $g(x)$, while translated left to get the graph of $k(x)$, down to get the graph of $m(x)$, and up to get the graph of $n(x)$. We normally think of a positive change to the domain as being further to the right and a positive change to the range as being further up. Given that convention, we have the apparent inconsistency. The graph of $p(x - 3)$ moves right, not left, and the graph of $p(x) + 3$ moves up as expected. It turns out that the apparent inconsistency is in the notation. We like to write our functions as $f(x) =$. Let's rewrite our functions in the form $y =$.

$g(x)$ stays $y = p(x - 3)$, $k(x)$ stays $y = p(x + 3)$ but $m(x)$ changes to $y + 3 = p(x)$ and $n(x)$ to $y - 3 = p(x)$.

Now we can have a consistent interpretation of the notation. The notation indicates the translation that is needed to bring the new graph *back* to the original graph. The graph of $g(x)$ must move left 3 units to get back to the graph of $p(x)$, so it must be 3 units to the right of $p(x)$. The graph of $n(x)$ must come down 3 units to get back to $p(x)$, so it must be 3 units above the graph of $p(x)$.

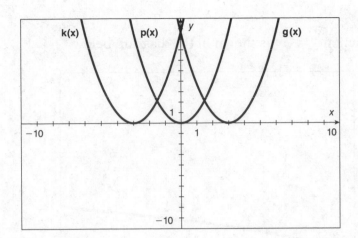 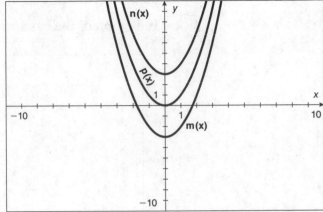

Describe the transformation for the function $g(x) = (x - 3)^2 + 1$

The graph moves to the right 3 units and up 1 unit. Use your graphing utility to verify this.

The graph of $y = ax^2$ is a stretch from the x-axis. It is important that you do not confuse the dilation from the origin that you studied in geometry (in which both the x- and y-coordinates are multiplied by the stretch factor) with a dilation from the x-axis (in which only the y-coordinate is multiplied by the stretch factor). If $0 < a < 1$ the graph moves closer to the x-axis, while if $a > 1$ the graphs moves farther from the x-axis. If $a < 0$, the graph is reflected over the x-axis.

Example: Describe the transformation of $y = x^2$ to get the graph of $p(x) = -3(x + 2)^2 + 4$.

Solution: An easy way to do this is to follow what happens to an input value of x. The first thing that happens is 2 is added to it (slide to the left 2), the value is squared (that is the function in question), the result is multiplied by -3 (reflect over the x-axis and stretch from the x-axis by a value of 3), and 4 is added (slide up 4).

Describe the transformation of $y = x^2$ to get the graph of $q(x) = \dfrac{1}{2}(x - 4)^2 - 3$.

The graph slides to the right 4, is stretched from the x-axis by a factor of $\dfrac{1}{2}$, and slides down 3.

The graph of the base function $y = \sqrt{x}$ is shown in the diagram below.

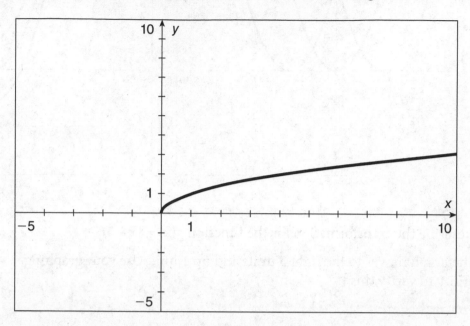

Describe the transformation of $y = \sqrt{x}$ needed to create the graph of $f(x) = 2\sqrt{x-3} - 1$.

The graph slides to the right 3, is stretched from the x-axis by a factor of 2, and slides down 1.

The graph of $y = |x|$ is shown below.

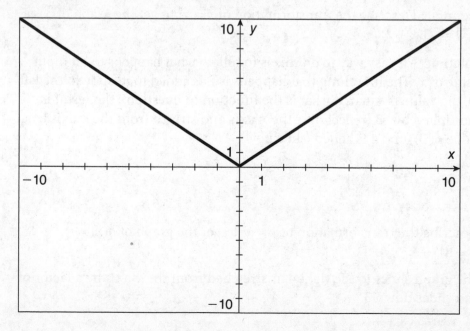

> Describe the transformation of $y = |x|$ needed to create the sketch of
> $y = -2|x + 3| + 1$

> The graph slides to the left 3 units, is reflected over the x-axis and
> stretched from the x-axis by a factor of 2, and moves up 1 unit.

EXERCISE 2.5

Describe the transformation of each of the base functions: $y = x^2$, $y = \sqrt{x}$, or $y = |x|$, whichever is appropriate.

1. $f(x) = \dfrac{-1}{2}(x+1)^2 - 3$

2. $g(x) = |x - 4| - 1$

3. $k(x) = \sqrt{x + 9} + 1$

4. $p(x) = -4x^2 + 9$

5. $q(x) = -2|x + 2| + 5$

6. The graph of $p(x)$ is shown below. Describe and sketch the graph of $y = 2p(x + 2) - 3$.

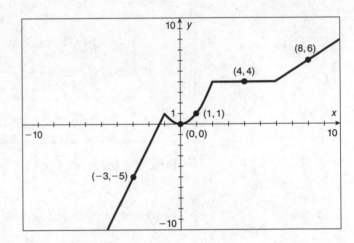

Inverse of a Function

To find the inverse of a function, the same notion of interchanging the x and y coordinates is applied. For example, to find the inverse of $f(x) = 2x - 7$, think about the function as $y = 2x - 7$. Switch the x and y: $x = 2y - 7$. Being as functions are written in the form $y =$ rather than $x =$, solve the equation for y. Add 7 to get $x + 7 = 2y$ and then divide by 2 to get $y = \dfrac{x+7}{2}$. If $f(x) = 2x - 7$ then $f^{-1}(x) = \dfrac{x+7}{2}$.

EXAMPLE

Find the inverse function of $g(x) = \dfrac{-3x+12}{5}$.

Rewrite the problem as $y = \dfrac{-3x+12}{5}$. Interchange the x and y to get $x = \dfrac{-3y+12}{5}$. Solve for y:

$5x = -3y + 12$ becomes $5x - 12 = -3y$ so $y = \dfrac{-5x+12}{3}$. The inverse of $g(x)$ is $g^{-1}(x) = \dfrac{-5x+12}{3}$.

EXAMPLE

Find the inverse of $p(x) = \dfrac{4x+9}{2x-5}$.

Rewrite the problem as $y = \dfrac{4x+9}{2x-5}$. Interchange the x and y to get $x = \dfrac{4y+9}{2y-5}$. Multiply both sides of the equation by $2y - 5$ to get $x(2y - 5) = 4y + 9$.

(Important: Remember that the goal is to solve for y. The work that follows is designed to meet this goal.)

Distribute:

$2xy - 5x = 4y + 9$

Gather terms in y on the left:

$2xy - 4y = 5x + 9$

Factor the y from the left side:

$y(2x - 4) = 5x + 9$

Solve for y:

$y = \dfrac{5x+9}{2x-4}$

The inverse of $p(x)$ is $p^{-1}(x) = \dfrac{5x+9}{2x-4}$.

EXERCISE 2.6

Find the inverse for each of the given functions.

1. Given $f(x) = -2x + 5$, find $f^{-1}(x)$

2. Given $g(x) = 4 - 5x$, find $g^{-1}(x)$

3. Given $k(x) = \dfrac{5x - 2}{x + 7}$, find $k^{-1}(x)$.

4. Given $v(x) = \dfrac{x - 10}{4x - 3}$, find $v^{-1}(x)$.

Graphical Representation of Functions

Sets of ordered pairs are useful for helping to clarify the concepts of relation, function, inverse, domain, and range, but, as you know, most of mathematics is done with formulas and graphs. By definition, a function is a relation in which no input value has multiple output values associated with it. What does that look like on a graph? It would mean that it would not be possible to draw a vertical line anywhere on the graph and have it hit more than one of the plotted points at any one time. (If the vertical line does not hit any of the points, that is fine. The requirement is that the vertical line cannot hit more than one point at a time.)

EXAMPLE

Which of the following graphs represent functions?

(a)

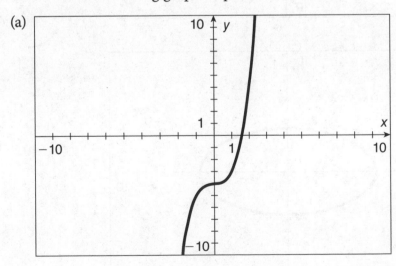

(b)

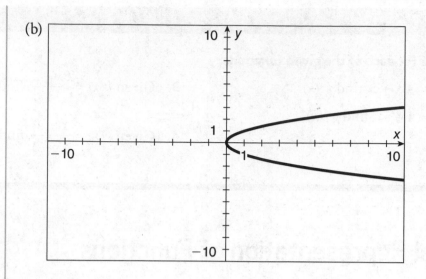

(c)

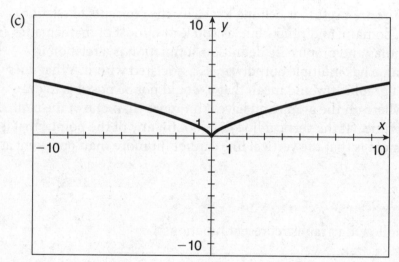

(d)

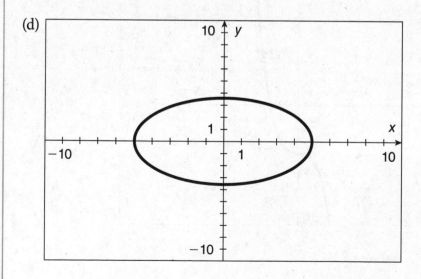

The graphs in choice (a) and choice (c) satisfy the vertical line test (a vertical line can never intersect each graph at more than one point) while the remaining two graphs fail to satisfy the vertical line test (it is possible for a vertical line to intersect each graph at more than one point).

At first, it is not as easy to determine if a relation represents a function when only given an equation. With experience, however, you will be able to tell which equations will probably not represent functions and which are likely to. For example, you most likely recognize that the equation $x^2 + y^2 = 36$ represents a circle with its center at the origin and a radius of 6. This is not a function. You also know that the equation $y = 4x^2$ is a parabola that opens up and has its vertex at the origin. This is a function. Do you know what the graphs of $x = 3y^2$ or $xy^3 - x^3y = 25$ look like? Neither is a function, and this can be shown by picking a value for x (e.g., $x = 1$) and noting that there is more than one value of y associated with it. Fortunately, these are not equations that will be encountered while studying Algebra 2.

Finding the inverse from a graph is not easy. Determining whether the graph of a relation is a function is not as difficult. Recall that the vertical line test is used to determine if a graph represents a function. If the inverse of the relation defined by the graph is to be a function, then none of the y coordinates can be repeated (if they were, then the graph of the inverse would fail the vertical line test). If the y coordinates cannot be repeated, then the graph would have to pass a horizontal line test. Given this, the relations represented by the graphs of (a) and (b) above have inverses that are functions.

To recap this important information:

- If a relation passes the vertical line test, the relation is a function.

- If a relation passes the horizontal line test, the inverse of the relation is a function.

Relations that pass both the vertical and horizontal lines tests are called **one-to-one** functions. You probably know that the inverse of $y = x^2$ is $\sqrt{x}$. However, based on the discussion above, this makes no sense. The graph of the parabola fails the horizontal line test, so the inverse does not exist. How do we resolve this conflict? We *restrict the domain* of the parabola from all real numbers to just the nonnegative numbers (that is, 0 and larger). The graph of this parabola is:

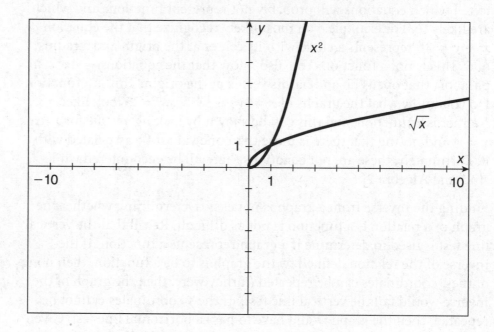

EXERCISE 2.7

Use these graphs to answer questions 1 and 2.

(a)

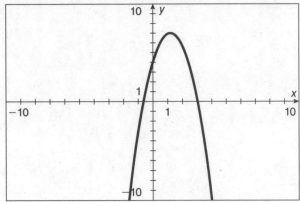

(c)

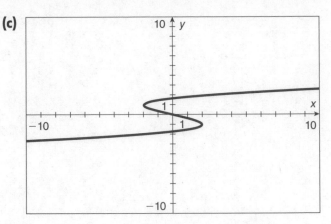

(b)

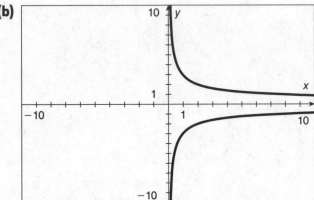

(d)

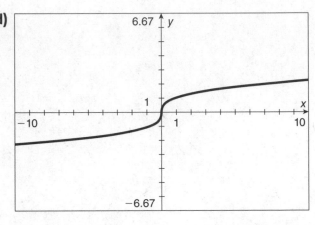

1. Which of the relations defined by the above graphs represents a function?

2. Which of the relations defined by the above graphs will have inverses that are functions?

Given the following representations, answer questions 3 and 4.

(a) $3x + y^2 = 4$ **(b)**

x	y
2	3
3	5
8	2
7	3
9	1
10	4

(c)

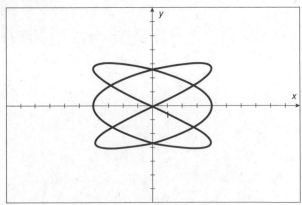

3. Which of the choices represent a function?

4. Which of the choices have an inverse that is a function?

CHAPTER 3

Quadratic Relationships

Equations of the form $y = mx + b$ are called linear relationships because the graph is a line. Linear relationships can also be expressed in the form $Ax + By = C$ and $\dfrac{x}{a} + \dfrac{y}{b} = 1$ (among others) because the exponent on the independent variable is 1. With quadratic relationships, the largest exponent in the equation is 2.

Special Factoring Formulas

There are a number of factoring "formulas" that you should recognize.

DIFFERENCE OF SQUARES

The expression $a^2 - b^2$ (two squares being subtracted) will always factor to $(a + b)(a - b)$.

> Factor $x^2 - 25$.
>
> $5^2 = 25$, so $x^2 - 25 = (x - 5)(x + 5)$

▶ Factor $16x^2 - 81y^4$.

▶ $16x^2 - 81y^4 = (4x)^2 - (9y^2)^2 = (4x - 9y^2)(4x + 9y^2)$

▶ Factor $75x^7 - 48xz^4$.

▶ Recall from Algebra I that the first step to factoring is to always remove common factors. You can see that $3x$ is a common factor for both terms in this expression, so $75x^7 - 48xz^4$ becomes $3x(25x^6 - 16z^4)$. This, in turn, becomes $3x\left(\left(5x^3\right)^2 - \left(4z^2\right)^2\right) = 3x\left(5x^3 - 4z^2\right)\left(5x^3 + 4z^2\right)$.

SQUARE TRINOMIALS

There are two forms to this style. $\left(a + b\right)^2 = a^2 + 2ab + b^2$ and $\left(a - b\right)^2 = a^2 - 2ab + b^2$. You start with a binomial (an expression with two terms) that is being squared. The first and last terms of the expansion are the squares of the terms forming the binomial (a^2 and b^2), while the middle term is twice the product of the terms forming the binomial. Once any common factors have been removed from the trinomial, we look at the first and third terms to determine if they are squares. If yes, then look at the middle term to see if it fits the pattern.

▶ Factor $4x^2 - 20x + 25$.

▶ The first and third terms are squares (the squares of $2x$ and 5) and the middle term is twice the product of $2x$ and 5 so $4x^2 - 20x + 25 = (2x - 5)^2$. (Observe that the sign between the terms in the binomial is the same as the sign of the middle term in the trinomial.)

▶ Factor $36x^2 + 168x + 196$.

▶ The common factor of 36, 168, and 196 is 4. Therefore, $36x^2 + 168x + 196$ becomes $4\left(9x^2 + 42x + 49\right) = 4\left((3x)^2 + 2(3x)(7) + (7)^2\right) = 4\left(3x + 7\right)^2$.

▶ Problems that appear to be complicated are often variations of some basic factoring pattern.

EXAMPLE

Factor $\left(4x^2-3\right)^2-44(4x^2-3)+484$

At first, it looks like you have to expand all terms to create a new polynomial with some pretty ugly terms in it. However, notice that the first term is clearly a square (since the exponent 2 is outside the parentheses) and that $484=22^2$. Consequently,

$$\left(4x^2-3\right)^2-44(4x^2-3)+484=\left(\left(4x^2-3\right)-22\right)^2=\left(4x^2-25\right)^2.$$

We're not done yet. The binomial $4x^2-25$ is the difference of squares.

$$\left(4x^2-25\right)^2=\left(\left(2x-5\right)\left(2x+5\right)\right)^2=\left(2x-5\right)^2\left(2x+5\right)^2.$$

Sum of Cubes: $\qquad a^3+b^3=\left(a+b\right)\left(a^2-ab+b^2\right)$

Difference of Cubes: $a^3-b^3=\left(a-b\right)\left(a^2+ab+b^2\right)$

Things you need to observe: the sign between the binomial in the factor is the same as the sign between the two cubes; the trinomial in the factor looks a great deal like a square trinomial *but* the significant difference is that the middle term is the product of the terms in the binomial (not twice the product) and the sign of the middle term is always the opposite of the sign between the binomials. You will be well served if you take the time to learn the cubes of the first 10 counting numbers as these are the most likely candidates you will use when working with cubes.

> Take the time to memorize these factoring patterns. Knowing them can save a great deal of time when solving equations.

EXAMPLE

Factor $27\,p^3-1,000$.

$$27\,p^3-1,000=\left(3p\right)^3-(10)^3=\left(3p-10\right)\left(\left(3p\right)^2+\left(3p\right)(10)+(10)^2\right)$$
$$=\left(3p-10\right)\left(9p^2+30p+100\right).$$

EXAMPLE

Factor $81p^3-3,000$

Neither 81 nor 3,000 is a cube, but they do have a common factor of 3. Therefore,

$$81p^3-3,000=3\left(27p^3-1,000\right)=3\left(3p-10\right)\left(9p^2+30p+100\right).$$

Factor $(2x-1)^3 + (3x-4)^3$.

Let's agree that this looks intimidating until we realize that it is not. Because $(2x-1)^3 + (3x-4)^3$ is the sum of two cubes, we simply need to follow the pattern to determine the factors.

$$(2x-1)^3 + (3x-4)^3 =$$
$$\big((2x-1)+(3x-4)\big)\big((2x-1)^2 - (2x-1)(3x-4) + (3x-4)^2\big)$$

$$(2x-1)^3 + (3x-4)^3 =$$
$$(5x-5)\big(4x^2 - 4x + 1 - (6x^2 - 11x + 4) + 9x^2 - 24x + 16\big)$$

$$(2x-1)^3 + (3x-4)^3 = 5(x-1)(7x^2 - 17x + 13)$$

Looking for patterns as a first step in the factoring process is the best way to approach the problem.

EXERCISE 3.1

Completely factor each of the following.

1. $4x^2 - 28x + 49$

2. $9x^2 + 30xy + 25y^2$

3. $100x^2 - 49y^2$

4. $9x^2 + 90xy + 225y^2$

5. $125x^3 - 512z^6$

6. $135c^3 + 40$

TRIAL AND ERROR

It is more common that the trinomial needing to be factored is not one of the special cases than that it is. It helps to remember that the product of two binomials $(ax + p)(bx + r) = abx^2 + (ar + bp)x + pr$.

Factor $45x^2 - 13x - 24$.

As you go through the trial and error process, keep in mind some basic facts. The constant at the end of the problem is a negative number. This means that the factors p and r must have different signs. The middle term is odd. This means that $ar + bp$, being odd, must have one term odd and the other even. So either both a and r are odd, or b and p are odd. Given this, let's take a look at the factors of 45 and 24.

45: 1, 3, 5, 9, 15, 45

24: 1, 2, 3, 4, 6, 8, 12, 24

EXAMPLE

▶ All the factors of 45 are odd, so we need one even and one odd factor for 24. The only choice for that is 3 and 8. What factors of 45 can be paired with 3 and 8 so that the term $3a + 8r = -13$? Note that $5*8 - 3*9 = 40 - 27 = 13$ gives us the correct number but the wrong sign. Therefore, we have the factors of $45x^2 - 13x - 24$ being $(9x - 8)$ and $(5x + 3)$.

▶ Factor $24x^2 - 74x + 45$

▶ The factors of 24 are 1, 2, 3, 4, 6, 8, 12, 24, and the factors of 45 are 1, 3, 5, 9, 15, 45. The last term is positive, so the signs in the factors must be the same, and the sign of the middle term is negative, which tells us that both signs are negative. The sum $ar + bp$, 74, is fairly large. Three (3) and 8 are not good choices for the factors of 24 because both factors of 45 are odd. This would cause one of the terms ar or bp to be the product of two odd numbers while the other would be the product of an odd and an even. Can you see how we cannot get 74 from these results? (Even times odd is even, odd times odd is odd, and the sum of an odd and an even is odd.) If we try the factors 1 and 24 with 1 and 45, we would get $(1)(24) + (1)(45) = 69$ or $(24)(45) + (1)(1) = 1,081$. These clearly do not work. To get a number in the neighborhood of 74 it makes sense to pick numbers from the middle of each list. You can gauge from here whether you will need to work with bigger or smaller values. Try 4 and 6 as the factors of 24 and 5 and 9 as the factors of 45. Pair these off to determine the value of $ar + bp$. One possibility is $(4)(9) + (6)(5) = 36 + 30 = 66$. Close, but not it. Switch the factors of 45 to get $(4)(5) + (6)(9) = 20 + 54 = 74$. Bingo! Therefore, $24x^2 - 74x + 45 = (4x - 9)(6x - 5)$.

We'll look at how to use your graphing calculator to get the factors a little later in this chapter.

EXERCISE 3.2

Completely factor each of the following.

1. $6x^2 - 13x + 6$

2. $20x^2 + x - 12$

3. $12x^2 + 44x - 45$

4. $60x^2 - 230x + 200$

5. $26x^2 + 57x - 20$

Completing the Square

We are going to learn how to change a quadratic expression into a form that resembles a square trinomial. The importance of this skill is that it enables us to look at the equations of circles and parabolas (and in later years of studies, other important relationships) in forms that tell us a good deal about the graph without actually having to sketch the graph.

We know that $4x^2 - 12x + 9$ can be written as $(2x - 3)^2$. Can you see how $4x^2 - 12x + 10$ can be written as $(2x - 3)^2 + 1$ or $4x^2 - 12x + 7$ can be written as $(2x - 3)^2 - 2$? Ten is 1 more than 9 so we add 1 to the square trinomial, while 7 is 2 less than 9 so we subtract 2 from the square trinomial.

Let's use the trinomials $ax^2 + bx + c$ and $4x^2 + 24x + 7$ to learn how to complete the square (trinomial).

Step 1. Factor the leading coefficient from the terms containing the variable.

From $ax^2 + bx + c$, factor the a to get $a\left(x^2 + \dfrac{b}{a}x\right) + c$ and from $4x^2 + 24x + 7$ factor the 4: $4\left(x^2 + 6x\right) + 7$.

Step 2. Take one-half the linear coefficient, square it, and add it inside the parentheses. Outside the parentheses, subtract the value added to the expression (*Don't* forget about the leading coefficient!)

Half of $\dfrac{b}{a}$ is $\dfrac{b}{2a}$. Therefore, $a\left(x^2 + \dfrac{b}{a}x\right) + c$ becomes:

$$a\left(x^2 + \frac{b}{a}x + \left(\frac{b}{2a}\right)^2\right) + c - a\left(\frac{b}{2a}\right)^2$$

while:

$$4\left(x^2 + 6x\right) + 7 \text{ becomes } 4\left(x^2 + 6x + (3)^2\right) + 7 - 4(3)^2.$$

Step 3. The trinomial inside the first set of parentheses is now a perfect square. Write it as such. Simplify the last part of the polynomial expression.

$$a\left(x^2 + \frac{b}{a}x + \left(\frac{b}{2a}\right)^2\right) + c - a\left(\frac{b}{2a}\right)^2 = a\left(x + \frac{b}{2a}\right)^2 + c - a\left(\frac{b^2}{4a^2}\right) =$$

$$a\left(x + \frac{b}{2a}\right)^2 + c - \frac{b^2}{4a} = a\left(x + \frac{b}{2a}\right)^2 + \frac{4ac - b^2}{4a}$$

Therefore, $ax^2 + bx + c = a\left(x + \dfrac{b}{2a}\right)^2 + \dfrac{4ac - b^2}{4a}$.

$$4\left(x^2 + 6x + (3)^2\right) + 7 - 4(3)^2 = 4\left(x + 3\right)^2 + 7 - 36 = 4\left(x + 3\right)^2 - 29.$$

Therefore, $4x^2 + 24x + 7 = 4\left(x + 3\right)^2 - 29$.

EXAMPLE

Complete the square on the quadratic $6x^2 + 8x - 7$.

Step 1:

$$6x^2 + 8x - 7 = 6\left(x^2 + \frac{8}{6}x\right) - 7 = 6\left(x^2 + \frac{4}{3}x\right) - 7$$

Step 2:

$$6\left(x^2 + \frac{4}{3}x\right) - 7 = 6\left(x^2 + \frac{4}{3}x + \left(\frac{2}{3}\right)^2\right) - 7 - 6\left(\frac{2}{3}\right)^2$$

Step 3:

$$6\left(x + \frac{2}{3}\right)^2 - 7 - \frac{24}{9} = 6\left(x + \frac{2}{3}\right)^2 - \frac{29}{3}$$

Therefore,

$$6x^2 + 8x - 7 = 6\left(x + \frac{2}{3}\right)^2 - \frac{29}{3}$$

(Yes, it is true that there are times when fractions cannot be avoided, but since you have a calculator that can do this arithmetic, don't get too worried about it.)

EXAMPLE

Complete the square on the trinomial $21x^2 - 84x + 25$.

Step 1:

$$21x^2 - 84x + 25 = 21\left(x^2 - 4x\right) + 25$$

Step 2:

$$21\left(x^2 - 4x\right) + 25 = -21\left(x^2 - 4x + (-2)^2\right) + 25 - 21(-2)^2$$

Step 3:

$$21\left(x^2 - 4x + (-2)^2\right) + 25 - 21(-2)^2 =$$
$$21\left(x^2 - 2\right)^2 + 25 - 84 = 21\left(x - 2\right)^2 - 59$$

Consequently,

$$21x^2 - 84x + 25 = 21\left(x^2 - 2\right)^2 - 59$$

EXERCISE 3.3

Complete the square for each quadratic expression.

1. $2x^2 + 12x + 5$

2. $2x^2 - 7x + 4$

3. $-5x^2 + 20x + 1$

4. $\frac{1}{2}x^2 - 3x - 9$

5. $\frac{2}{3}x^2 - 12x - 11$

Quadratic Formula

You learned in Algebra I about the **Zero Product Rule**: If the product of a and b is 0, then either $a = 0$ or $b = 0$ (or both could be 0). You extended this to solve quadratic equations. If $(x - 4)(x + 3) = 0$ than either $x = 4$ or $x = -3$. This works well when you are able to factor the quadratic expression. What happens when you cannot factor the expression?

As we saw in the last section, we can rewrite quadratic expressions to resemble square trinomials. For example, we saw that $4x^2 + 24x + 7$ can be written as $4(x+3)^2 - 29$. We can use this to solve the corresponding quadratic equation.

EXAMPLE

Solve $4x^2 + 24x + 7 = 0$

$$4x^2 + 24x + 7 = 4(x+3)^2 - 29.$$

Therefore, $4x^2 + 24x + 7 = 0$ becomes $4(x+3)^2 - 29 = 0$. Add 29 to both sides of the equation: $4(x+3)^2 = 29$. Take the square root of both sides of the equation (remembering to use both the positive and negative terms):

$2(x+3) = \pm\sqrt{29}$. Divide by 2 and subtract 3 to get $x = -3 \pm \dfrac{\sqrt{29}}{2}$.

As nice as that is, it would take a great deal of time to complete the square every time we had a quadratic equation that could not be factored. The **quadratic formula** is the result of completing the square on the general quadratic equation $ax^2 + bx + c = 0$. Using the result of the previous section, this equation becomes:

$$a\left(x + \frac{b}{2a}\right)^2 + \frac{4ac - b^2}{4a} = 0.$$

Add the constant to the right: $a\left(x + \dfrac{b}{2a}\right)^2 = \dfrac{b^2 - 4ac}{4a}$. (Did you see how the numerator was negated?)

Divide by a:

$$\left(x + \frac{b}{2a}\right)^2 = \frac{b^2 - 4ac}{4a^2}$$

Take the square root of both sides of the equation:

$$x + \frac{b}{2a} = \pm\sqrt{\frac{b^2 - 4ac}{4a^2}} = \pm\frac{\sqrt{b^2 - 4ac}}{2a}$$

Solve for x:

$$x = \frac{-b \pm \sqrt{b^2 - 4ac}}{2a}.$$

EXAMPLE

Use the quadratic formula to solve $4x^2 + 24x + 7 = 0$.

Solving,

$$x = \frac{-24 \pm \sqrt{24^2 - 4(4)(7)}}{2(4)} = \frac{-24 \pm \sqrt{576 - 112}}{8} = \frac{-24 \pm \sqrt{464}}{8} =$$

$$\frac{-24 \pm \sqrt{(16)(29)}}{8} = \frac{-24}{8} \pm \frac{4\sqrt{29}}{8} = -3 \pm \frac{\sqrt{29}}{2}.$$

This is certainly a faster process!

EXAMPLE

Use the quadratic formula to solve $6x^2 + 8x - 7 = 0$.

Solving,

$$x = \frac{-8 \pm \sqrt{8^2 - 4(6)(-7)}}{2(6)} = \frac{-8 \pm \sqrt{64 + 168}}{12} = \frac{-8 \pm \sqrt{232}}{12} = \frac{-8}{12} \pm \frac{\sqrt{4}\sqrt{58}}{12} = \frac{-2}{3} \pm \frac{\sqrt{58}}{6}$$

EXAMPLE

Use the quadratic formula to solve $\frac{-1}{2}x^2 - 3x + 5 = 0$.

Solving,

$$x = \frac{-(-3) \pm \sqrt{(-3)^2 - 4\left(\frac{-1}{2}\right)(5)}}{2\left(\frac{-1}{2}\right)} = \frac{3 \pm \sqrt{9 + 10}}{-1} = -3 \pm \sqrt{19}$$

EXERCISE 3.4

Solve each of the following quadratic equations.

1. $9x^2 - 11x - 2 = 0$

2. $-14x^2 + 27x + 11 = 0$

3. $4x^2 + 25x + 6 = 0$

4. $4x^2 + 25x - 6 = 0$

5. $12x^2 + 13x - 3 = 0$

The Parabola

The most basic quadratic relationship is the parabola. The standard form of the equation is $y = f(x) = ax^2 + bx + c$. The axis of symmetry for the parabola in standard form is $x = \dfrac{-b}{2a}$ and the coordinates of the vertex in standard form are $\left(\dfrac{-b}{2a}, f\left(\dfrac{-b}{2a} \right) \right)$.

Example: Find the coordinates of the vertex of the parabola $f(x) = 2x^2 - 4x + 5$.

Solution: The axis of symmetry for the parabola is $x = 1$. The y-coordinate of the vertex is $f(1) = 2(1)^2 - 4(1) + 5 = 3$. Therefore, the vertex is at $(1, 3)$.

Use the technique of completing the square on $f(x) = 2x^2 - 4x + 5$, leaving your answer in the form $f(x) = a(x - h)^2 + k$.

Solving,

$$f(x) = 2x^2 - 4x + 5$$
$$= 2\left(x^2 - 2x\right) + 5$$
$$= 2\left(x^2 - 2x + (-1)^2\right) + 5 - 2(-1)^2$$
$$= 2\left(x - 1\right)^2 + 3$$

Notice that this form of the equation identifies the vertex of the parabola, $(1, 3)$. The vertex form of the parabola is $f(x) = a(x - h)^2 + k$. The axis of symmetry is $x = h$ and the coordinates of the vertex of the parabola are (h, k).

EXAMPLE

▶ Find the coordinates of the vertex of the parabola $f(x) = -4(x + 5)^2 + 3$.

▶ The vertex of the parabola is $(-5, 3)$.

EXERCISE 3.5

Find the equation of the axis of symmetry and the coordinates of the vertex for the parabola described.

1. $y = 3x^2 + 12x - 4$

2. $f(x) = 4(x + 5)^2 + 2$

3. $g(x) = -2x^2 + 3x + 4$

4. $p(x) = \dfrac{-2}{3}(x - 3)^2 + 6$

5. $q(x) = \dfrac{-2}{9}x^2 + \dfrac{8}{3}x + 2$

Applications

The leading coefficient of the parabola $y = ax^2 + bx + c$ or $y = a(x - h)^2 + k$ determines whether the graph of the parabola is concave up (opens up) or down. If the graph is concave up, the y-coordinate of the vertex is the minimum value of the function, while if the graph is concave down, the y-coordinate of the vertex is the maximum value of the domain.

EXAMPLE

▶ A manufacturer determines that the weekly cost function for producing n items of her product is $C(n) = 20n + 1{,}200$. She has also determined that the price, p, she can charge for each of these n items is given by $p = 770 - 25n$ dollars. How many units must she produce each week to maximize her profit?

▶ Profit is the difference between Revenue, R, and Cost. Her Revenue is the product of the price per unit and the number of units she sells. That is, $R = np = n(770 - 25n) = 770n - 25n^2$. Her profit function is $P(n) = 770n - 25n^2 - (20n + 1{,}500) = -25n^2 + 750n - 1{,}500$. The vertex for the associated parabola occurs when $n = \dfrac{-750}{2(-25)} = 15$ items. Her weekly profit for these 15 items is \$4,125.

A homeowner has 400 feet of fencing to use to enclose a rectangular garden. One side of the garden lies along a creek with straight shoreline (the miracles of modern mathematics!). The homeowner wishes to know the dimensions of the garden with maximum area. What are they?

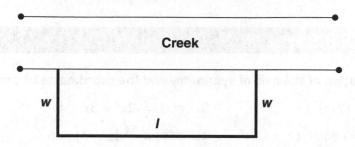

The area of the rectangle is $A = lw$ and the number of feet of fencing used can be expressed by the equation $l + 2w = 400$ or $l = 400 - 2w$. Substitute into the area equation to get $A = 400w - 2w^2$. The axis of symmetry for the corresponding parabola is $w = \dfrac{-400}{-4} = 100$. The width of the garden plot is 100 feet while the length of the plot will be 200 feet giving the garden an area of 20,000 square feet.

Drawing a diagram of the problem often helps you see how the variables of a problem fit together.

An amateur golfer tees up a golf ball and hits the ball with an approximate speed of 85 miles per hour. He elevates the ball at an angle of 41 degrees. The law of vectors tells us that the vertical position of the ball t seconds after it is hit is given by the equation $v = -16t^2 + 80.1t$ and the horizontal position of the ball is given by the equation $h = 95.5t$ (with distance measured in feet). (Since the height of the ball when it is on a tee is negligible compared to the magnitude of the other numbers, we are assuming an initial height of 0.)

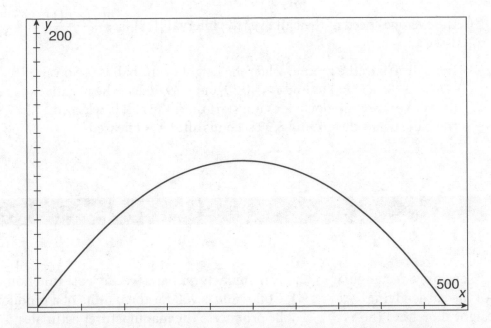

▶ What is the highest position of the ball? When does the ball strike the ground? How far down the course is the ball when it stops (measure in yards) assuming the ball travels another 20 yards after it hits the ground?

▶ The ball is at its maximum height when $t = \dfrac{-80.1}{-32} = 2.5$ seconds. The height of the ball is 100 feet. Solve $-16t^2 + 80.1t = 0$ to determine that the ball lands after 5.01 seconds. The distance the ball has traveled down the fairway is $\dfrac{95.5 * 5.01}{3} = 159.5$ yards. Adding 20 yards for the roll gives a drive of 179.5 yards.

EXAMPLE

▶ A ball is thrown vertically into the air from the edge of the roof of a building. The height of the ball, in feet, t seconds after it is thrown is given by the equation $h(t) = -16t^2 + 128t + 52$.

(a) What is the maximum height of the ball?

(b) What is the average speed of the ball from $t = 1$ to $t = 3$ seconds?

(c) At what time does the ball strike the ground?

▶ Solutions:

(a) The ball reaches its maximum height at $t = \dfrac{-128}{2(-16)} = 4$ seconds.

The height of the ball is $h(4) = -16(4)^2 + 128(4) + 52 = 308$ ft.

(b) The average speed of the ball over the interval [1, 3] is $\dfrac{h(3)-h(1)}{3-1}=64$ ft/sec.

(c) The ball strikes the ground when the height of the ball is 0. Solve $-16t^2+128t+52=0$ to find $t=-0.39$ and 8.39. Reject the negative answer because the clock does not start until the ball is released. The ball strikes the ground 8.39 seconds after it is released.

EXERCISE 3.6

Answer the following questions.

1. A homeowner wishes to enclose a rectangular garden with one side of the garden lying along a creek with straight shoreline. The homeowner wishes to know the dimensions of the garden with maximum area that can be enclosed for a total of $1,000. The fencing alongside the creek needs to be sturdier than the rest of the fencing. As a consequence, the fencing alongside the creek cost $20 per foot while remaining fence costs $8 per foot. What are the dimensions of the garden with a maximum area that can be enclosed?

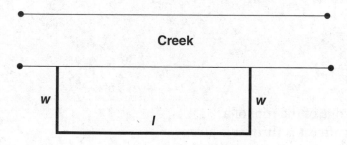

2. A manufacturer has a weekly cost function $C(n) = 80n + 900$ for the n units of a product he produces. The manufacturer estimates that the weekly demand for his product at a price of p is $p = 150 - 0.5n$. Determine the maximum profit the manufacturer can make under these conditions.

Use this information to answer questions 3–5.

A ball is thrown vertically in the air with an initial velocity of 112 feet per second from the top edge of an 80-foot building. The height of the ball after t seconds is given by the equation $h = -16t^2 + 112t + 80$.

3. What is the maximum height of the ball?

4. What is the average speed of the ball on the interval [1, 3]?

5. At what time does the ball strike the ground?

Use this information for questions 6–8.

An amateur golfer tees up a golf ball and hits the ball with an approximate speed of 100 miles per hour. He elevates the ball at an angle of 37 degrees. The law of vectors tells us that the vertical position of the ball t seconds after it is hit is given by the equation $v = -16t^2 + 88.3t$ and the horizontal position of the ball is given by the equation $h = 117.1t$ (with distance measured in feet). (Since the height of the ball when it is on a tee is negligible compared to the magnitude of the other numbers, assume an initial height of 0.)

6. What is the highest position of the ball?

7. When does the ball strike the ground?

8. How far down the course is the ball when it stops (measure in yards) assuming the ball travels another 20 yards after it hits the ground?

Using the Parabola to Factor

Here are three questions that ask for the same information:

Find the roots of the equation $x^2 - x - 6 = 0$.

Determine the values of the x-intercepts for the graph of $f(x) = x^2 - x - 6$.

Determine the zeroes of the function $f(x) = x^2 - x - 6$.

In each case, the values of x are –2 and 3. The x-intercepts are the points on the graph when the y-coordinate is 0, and the zeroes of a function are those values of x for which the output is 0. We can use this information to factor quadratic expressions. (A word of caution here first. The difference between an expression and an equation is the presence of an equal sign. $x^2 - x - 6 = 0$ is a quadratic equation, while $x^2 - x - 6$ is a quadratic expression. We are going to use equations to answer questions about expressions.)

Given the function $f(x) = x^2 - 5x - 24$, the graph of this function is shown.

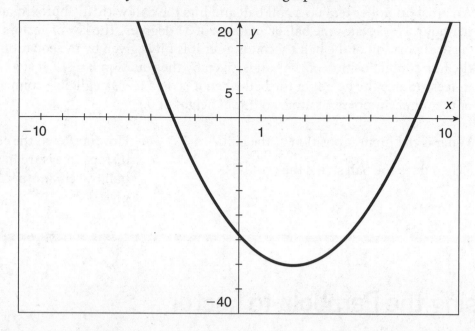

Use this example to practice with your graphing calculator how to find zeros of a function.

You can see that the graph crosses the x-axis at $x = -3$ and $x = 8$. Working with these equations, we get $x + 3 = 0$ and $x - 8 = 0$. The Zero Product Property allows us to write $(x + 3)(x - 8) = 0$. Therefore, $x^2 - 5x - 24 = (x + 3)(x - 8)$. This might not seem all that impressive because it is more than likely that you would have no trouble factoring $x^2 - 5x - 24$.

Let's try something a little more challenging.

EXAMPLE

Factor $48x^2 + 64x - 35$. Sketch the graph of $y = 48x^2 + 64x - 35$ on your graphing calculator using a window that clearly shows the x-intercepts.

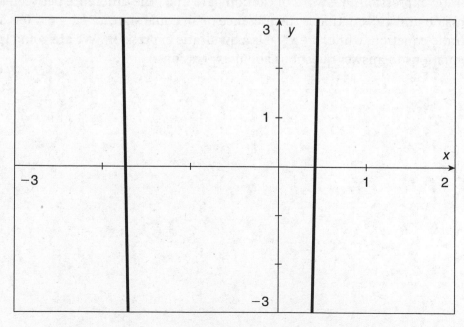

Use the Zero feature on your calculator to find:

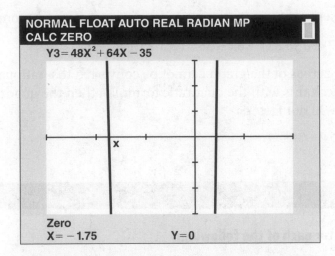

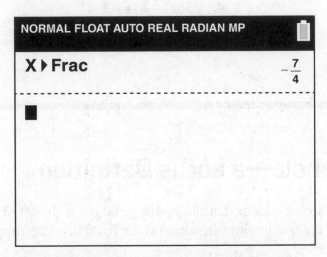

We now know that $y = 0$ when $x = \dfrac{-7}{4}$ and when $x = \dfrac{5}{12}$. Manipulating these equations we get $4x = -7$ and $12x = 5$ so that $4x + 7 = 0$ and $12x - 5 = 0$. We now have $48x^2 + 64x - 35 = (4x + 7)(12x - 5)$.

Use the graph of $y = 8x^2 - 6x - 35$ to factor $8x^2 - 6x - 35$.

The x-intercepts of the graph are $x = \dfrac{5}{2}$ and $x = \dfrac{-7}{4}$. These become $2x - 5 = 0$ and $4x + 7 = 0$ so $8x^2 - 6x - 35 = (2x - 5)(4x + 7)$.

Note: If the zeroes of the graph cannot be converted to a rational number (you can check this with the quadratic formula), then the quadratic expression will not factor.

EXERCISE 3.7

Use a graph or an equation to help factor each of the following.

1. $72x^2 + 42x - 85$

2. $72x^2 - 175x + 98$

3. $180x^2 - 9x - 56$

4. $350x^2 - 41x - 171$

5. $12x^2 - 25x - 323$

The Parabola—a Locus Definition

The parabola also has a locus definition. The parabola is the set of all points equidistant from a fixed point (the focus) and a fixed line (the directrix).

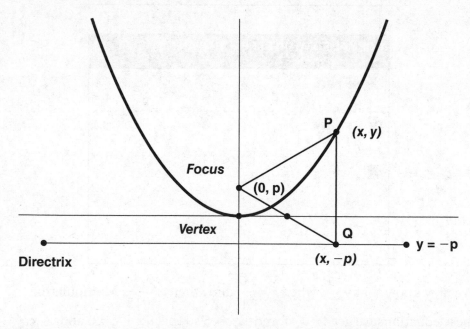

By definition, $PQ = PFocus$ so that $y + p = \sqrt{x^2 + (y - p)^2}$. Square both sides of the equation: $(y + p)^2 = x^2 + (y - p)^2$ or $y^2 + 2py + p^2 = x^2 + y^2 - 2py + p^2$ so that $4py = x^2$ and, finally, $y = \dfrac{1}{4p} x^2$.

Notice that the distance from the focus to the vertex is p as is the distance from the vertex to the directrix. This parabola has its vertex at the origin. We already saw that the equation of the parabola with vertex off the origin has its equation $y = a(x - h)^2 + k$. We see that the coefficient gives a statement about the location of the focus and directrix. We also get to examine the graphs of $x = a(y - k)^2 + h$, the parabola that opens to the right (when $a > 0$), and to the left (when $a < 0$).

EXAMPLE

▶ Determine the coordinates of the focus and equation of the directrix for the parabola $y = x^2$.

▶ With $a = 1$, we set $\dfrac{1}{4p} = 1$ and determine that $p = \dfrac{1}{4}$. Therefore, the focus is located at $\left(0, \dfrac{1}{4}\right)$ and the directrix has equation $y = \dfrac{-1}{4}$.

EXAMPLE

▶ The directrix of a parabola has equation $y = 3$ and the focus has coordinates (2, –1). What is the equation of the parabola?

▶ The parabola always opens toward the focus, so in this case it opens down. The coordinates of the vertex are (2, 1) because the vertex is always midway between the focus and directrix. Finally, p is the distance from the vertex to the focus (or the vertex to the directrix), so $p = 2$. The equation of the parabola is $y = \dfrac{-1}{8}(x - 2)^2 + 1$.

EXAMPLE

▶ The equation for the directrix of a parabola is $x = 3$ and the coordinates of the vertex are (7, 3). What is the equation of the parabola?

▶ Since the directrix is a vertical line, the equation of the parabola will be $x = \dfrac{1}{4p}(y - 3)^2 + 7$. The distance from the vertex to the directrix is 4, so $4p = 16$. Therefore, the equation is $x = \dfrac{1}{16}(y - 3)^2 + 7$.

EXERCISE 3.8

1. Find the equation of the directrix and coordinates of the focus for the parabola $y = 8(x+2)^2 - 1$.

2. Find the equation of the parabola that has a focus at (2, 5) and a vertex at (2, 9).

3. Find an equation for the parabola that has a focus at (2, 5) and a directrix at $y = 11$.

4. Find an equation for the parabola with directrix at $x = -4$ and focus at (8, 3).

5. Find the equation of the directrix for the parabola with equation $x = \dfrac{-1}{32}(y-8)^2 - 1$.

Factoring by Grouping

You have been using the distributive property for at least five years now. Knowing that $a(b + c) = ab + ac$ is really the basis for all factoring. For example, $x^2 + 5x + 6 = x^2 + 2x + 3x + 6 = (x^2 + 2x) + (3x + 6) = x(x + 2) + 3(x + 2)$. Notice that the binomial $x + 2$ serves as the "a" in the distributive property and this allows us to remove the common factor to get $(x + 2)(x + 3)$.

There are certain cubic and higher order polynomials that can be factored in a similar style. For example, $x^3 + 3x^2 - x - 3 = (x^3 + 3x^2) - (x + 3) = x^2(x+3) - (x+3)$. With $x + 3$ as a common factor, $x^2(x+3) - (x+3) = (x+3)(x^2 - 1) = (x+3)(x+1)(x-1)$. It is really just a matter of recognizing that a binomial can be a common factor.

EXAMPLE

Factor $x^3 - 5x^2 + 4x - 20$.

Gather the first and second terms together as well as the third and fourth, $(x^3 - 5x^2) + (4x - 20)$. Take out the common factor in each grouping, $x^2(x-5) + 4(x-5)$. Notice that the binomial $x - 5$ is a common factor. Remove it, $(x-5)(x^2 + 4)$.

EXAMPLE

▶ Factor $x^3 + 4x^2 - 25x - 100$.

▶ Gather 2 by 2, $(x^3 + 4x^2) - (25x + 100)$. (Be careful about the second term because of the negative sign outside the parentheses.) Remove the common factor for each grouping, $x^2(x+4) - 25(x+4)$. Remove the common binomial factor, $(x+4)(x^2 - 25)$. The second factor is the familiar difference of squares. Finish by factoring this, $(x+4)(x-5)(x+5)$.

EXAMPLE

▶ Factor $x^4 + 4x^3 - 125x - 500$.

▶ This problem certainly looks different than the others, but the process is the same. Group 2 by 2, $(x^4 + 4x^3) - (125x + 500)$. Remove the common factors for each group, $x^3(x+4) - 125(x+4)$. Remove the common binomial factor, $(x+4)(x^3 - 125)$. The second factor in this case is the difference of cubes. Finish by factoring this, $(x+4)(x-5)(x^2 + 5x + 25)$.

Not all factoring problems are this easy to recognize. These types of problems are a special case—ones that you should recognize. We will look at more "challenging" problems later in the book.

EXERCISE 3.9

Completely factor each of the following.

1. $x^3 + 4x^2 - 25x - 100$

2. $x^4 + 4x^3 - 125x - 500$

3. $x^3 - 7x^2 - 81x + 567$

4. $x^4 + 3x^3 + 27x + 81$

5. $x^3 - 5x^2 + 27x - 135$

EXERCISE 3.10

You've had the advantage of knowing what type of problem you were factoring based on the section heading. Here is a mixture of problems for you to practice identifying the patterns involved.

1. $5x^3 - 405x$

2. $4x^2 - 36xy + 81y^2$

3. $20x^2 - 33x - 27$

4. $154x^3 + 69x^2 - 108x$

5. $x^3 - 8x^2 - 16x + 128$

6. $x^4 - 10x^2 + 25$

7. $x^4 - 50x^2 + 625$

8. $x^4 + 4x^3 - 1,000x - 4,000$

9. $15x^2 - 34x + 15$

10. $48x^3 + 53x^2 - 45x$

Circles

A circle is the set of coplanar points in a plane that are at a fixed distance (radius) from a fixed point (the center). If (h, k) represent the coordinates of the center and r is the length of the radius of the circle, the equation becomes $(x - h)^2 + (y - k)^2 = r^2$.

EXAMPLE

> Determine the coordinates of the center and the length of the radius of a circle with equation $x^2 + y^2 - 10x + 8y - 23 = 0$.

> This is a classic example of how the technique of completing the square is used for something other than solving an equation. Gather the terms in x together as well as those in y, moving the constant to the right side of the equation, $x^2 - 10x + y^2 + 8y = 23$. Completing the square in each variable: $x^2 - 10x + 25 + y^2 + 8y + 16 = 23 + 25 + 16$. Factor terms: $(x - 5)^2 + (y + 4)^2 = 64$. The center of the circle is (5, –4) and the radius has length 8.

▶ The endpoints of a diameter of a circle have coordinates (–3, 5) and (9, –1). Write the equation of the circle.

▶ The midpoint of the diameter is the center of the circle. The coordinates of the center are $\left(\dfrac{-3+9}{2}, \dfrac{5+(-1)}{2}\right) = (3, 2)$. The length of the radius of the circle is $\sqrt{(-3-3)^2 + (2-5)^2} = \sqrt{45}$. Therefore, the equation of the circle is $(x-3)^2 + (y-2)^2 = 45$.

▶ Determine the coordinates of the intersection of the circles $x^2 + y^2 = 50$ and $(x-3)^2 + y^2 = 35$.

▶ Rewrite $x^2 + y^2 = 50$ as $y^2 = 50 - x^2$ and substitute this into the second equation: $(x-3)^2 + 50 - x^2 = 35$. Expand and solve:

$$x^2 - 6x + 9 + 50 - x^2 = 35$$

$$-6x + 59 = 35$$

$$6x = 24$$

$$x = 4$$

▶ When $x = 4$, $16 + y^2 = 50$ so $y^2 = 34$ and $y = \pm\sqrt{34}$. Therefore, the coordinates of intersection are $\left(4, \sqrt{34}\right)$ and $\left(4, -\sqrt{34}\right)$.

▶ Determine the coordinates of intersection of the circle $x^2 + y^2 = 10$ and $y = x^2 - 8$.

▶ Substitute $x^2 - 8$ for y and rewrite the equation of the circle as $x^2 + \left(x^2 - 8\right)^2 = 10$. Expand and solve:

$$x^2 + x^4 - 16x^2 + 64 = 10$$

$$x^4 - 15x^2 + 54 = 0$$

$$\left(x^2 - 9\right)\left(x^2 - 6\right) = 0$$

$$x = \pm 3, \pm\sqrt{6}$$

▶ If $x = -3$, $y = 9 - 8 = 1$. We get the same result when $x = 3$. When $x = \sqrt{6}$, $y = 6 - 8 = -2$. We get the same result when $x = -\sqrt{6}$. (Warning: if we had substituted the values of x into the equation of the circle, we would have gotten extraneous roots. For example, when $x = -3$, $y = \pm 1$. The point $(-3, -1)$ is not on the parabola. Be careful.) The solution to the problem is $(-3, 1)$, $(3, 1)$, $(\sqrt{6}, -2)$, and $(-\sqrt{6}, -2)$

EXERCISE 3.11

Answer the following questions.

1. Determine the coordinates of the center and length of the radius of the circle with equation $x^2 + y^2 - 12x + 8y = 144$.

2. Write the equation of the circle with center at $(-4, 5)$ and is tangent to the x-axis.

3. Write the equation of the circle that has a diameter with endpoints at $(-3, 10)$ and $(7, 22)$.

4. Find the coordinates of the intersection of the circles $x^2 + y^2 = 45$ and $x^2 + (y + 3)^2 = 66$

5. Determine the coordinates of the points of intersection between the circle $x^2 + y^2 = 25$ and the parabola $y = \dfrac{-1}{2}x^2 - \dfrac{3}{2}x + 5$.

CHAPTER 4

Complex Numbers

Since you were in elementary school you have been learning about sets of numbers. Some of the names are really perfect. The **counting numbers**—well, hold your fingers out and start counting. Soon after, you learned about the **rational numbers**. It wasn't that these numbers were sane and made sense, but because they came from ratios. If some numbers were not rational, then there must be the **irrational numbers**—those that were not rational. Arguments can be made about whether the **whole numbers** and **integers** translate well into English, but that is not important at this time. The one name that everyone thought made sense is the **real numbers**. You can "see" real numbers. Count your fingers, break a cookie into pieces and describe what part of the original cookie each is, look at the hypotenuse of a right triangle whose legs have length 2 and 3.

The problem is, it's all wrong. There is no such thing as a 1. You can hold up a finger or a pencil. You have a finger and a pencil. That the quantity is represented by the word *one* is a linguistic convention. (This argument is not different from what is red.) At the time these terms were coined, they made sense to people. Real numbers were numbers that could be put on a number line. However, Gerolamo Cardano (1501–1576) and other sixteenth-century mathematicians began studying a new set of numbers that could *not* be put on a number line. "Well," we might assume that they thought, "if these numbers aren't real, they must be **imaginary**."

Powers of *i*

In Algebra I you learned that the solutions to the equation $x^2 = 1$ were $x = \pm 1$, while the solution to $x^2 = -1$ didn't exist (whereas the true answer at that point in time should have been that it did not exist within the set of *real* numbers). It was in the sixteenth century that mathematicians began looking into the possibility that there were solutions to this problem. They introduced the number *i* to represent the positive solution. It is by definition that we claim

$\sqrt{-1} = i$. The arithmetic rules were exactly the same as those for real numbers. If $\sqrt{3} + \sqrt{3} = 2\sqrt{3}$, then:

$$i + i = 2i. \text{ If } \left(\sqrt{3}\right)^2 = 3 \text{ , then } \left(\sqrt{-1}\right)^2 = i^2 = -1.$$

The powers of i turn out to repeat themselves in a cycle of four steps: $i^2 = -1$; $i^3 = i^2 i = -i$; $i^4 = (i^2)^2 = (-1)^2 = 1$. The next four powers of i simply remove $i^4 = 1$. That is, $i^5 = i^4 i = i$; $i^6 = i^4 i^2 = -1$, $i^7 = i^4 i^3 = -i$, and $i^8 = i^4 i^4 = 1$. In general, to evaluate i^n, divide n by 4 and use the remainder of that division to get your answer (realizing that $i^0 = 1$ because any nonzero number raised to the zero power is 1).

Evaluate i^{273}

$273 \div 4 = 68$ with a remainder of 1. Therefore, $i^{273} = i$.

Evaluate i^{-73}.

$i^{-73} = \dfrac{1}{i^{73}}$ (definition of a negative exponent). We'll multiply the numerator and denominator by i^3 because this will make the exponent in the denominator a multiple of 4 and we know that:

$$i^4 = 1, \left(\frac{1}{i^{73}}\right)\left(\frac{i^3}{i^3}\right) = \frac{i^3}{1} = -i.$$

EXERCISE 4.1

Evaluate each of the following.

1. i^{38}

2. i^{138}

3. i^{-57}

Simplifying Imaginary Numbers

Evaluating the square root of a negative number requires that you first factor $\sqrt{-1}$ from the expression and then proceed as you would with irrational numbers.

Working with imaginary numbers is exactly the same as working with irrational numbers. You just need to remember that $i = \sqrt{-1}$.

EXAMPLE

▶ Simplify $\sqrt{-27}$

$$\sqrt{-27} = \sqrt{-1}\,\sqrt{27} = i\,\sqrt{9}\,\sqrt{3} = 3i\sqrt{3}$$

EXAMPLE

▶ Simplify $5\sqrt{-12} + 8\sqrt{-27}$.

▶ Remember, first deal with the square root of –1 and then continue on as normal.

$$5\sqrt{-12} + 8\sqrt{-27} =$$
$$5\sqrt{-1}\,\sqrt{(4)(3)} + 8\sqrt{-1}\,\sqrt{(9)(3)} = 10i\sqrt{3} + 24i\sqrt{3} = 34i\sqrt{3}.$$

EXAMPLE

▶ Simplify $\left(4\sqrt{-8}\right)\left(7\sqrt{-18}\right)$

▶ It is particularly important to first factor the square root of –1 before doing anything else.

$$\left(4\sqrt{-8}\right)\left(7\sqrt{-18}\right) = \left(4\sqrt{-1}\sqrt{4}\sqrt{2}\right)\left(7\sqrt{-1}\sqrt{9}\sqrt{2}\right) =$$
$$28(-1)(6)(2) = -336$$

▶ (Had you not first factored the square root of –1, you would have gotten 336 as your answer, and that is not correct.).

EXAMPLE

▶ The problems illustrated so far can all be done on your graphing calculator. In this example we'll include literal values. (Literal values are represented by letters but are not variables. In the equation $y = mx + b$, m and b are literal values while y and x are variables.) For the sake of simplicity, assume each literal value is a positive integer.

▶ Simplify $17\sqrt{-28c} - 9\sqrt{-63c}$.

$$17\sqrt{-28c} - 9\sqrt{-63c} = 17\left(\sqrt{-1}\right)\left(\sqrt{4}\right)\left(\sqrt{7c}\right) - (9)\left(\sqrt{-1}\right)\left(\sqrt{9}\right)\left(\sqrt{7c}\right) =$$
$$34i\sqrt{7c} - 27i\sqrt{7c} = 7i\sqrt{7c}.$$

EXERCISE 4.2

Simplify each of the following. (Assume any literal value to be a positive integer.)

1. $4\sqrt{-50} - 6\sqrt{-32}$

2. $17\sqrt{-24} + 5\sqrt{-54}$

3. $\left(\sqrt{-72}\right)^2$

4. $\left(12\sqrt{-20}\right)\left(20\sqrt{-12}\right)$

5. $\dfrac{52\sqrt{-18}}{13\sqrt{-12}}$

6. $\left(40\sqrt{-45a}\right)\left(15\sqrt{-20a}\right)$

7. $\dfrac{40\sqrt{-45a^3}}{15\sqrt{-20a}}$

Arithmetic of Complex Numbers

The arithmetic rules for imaginary numbers are exactly the same as the rules for irrational numbers, as they should be. Remember, i is defined by a square root. However, we also work with a combination of integers and irrational numbers. You can add $4 + 3\sqrt{5}$. The rules for the combination of a real and imaginary component will behave in the same way. However, we need to realize that we have created an entirely new set of numbers in the process. They are not real and they are not imaginary, so this must be really hard to think about because it is so complex. Yes, we now have the set of complex numbers. These numbers take the form $a + bi$ where both a and b are real numbers. Complex numbers have a real component, a, and an imaginary component, bi. If $a = 0$, the number is imaginary, and if $b = 0$, the number is real. Therefore, the real numbers and imaginary numbers are subsets of the complex numbers.

Addition and subtraction are done in the familiar manner that the real components are added together as are the imaginary components, $\left(8 + 7i\right) + \left(5 - 3i\right) = 13 + 4i$ and $\left(8 + 7i\right) - \left(5 - 3i\right) = 3 + 10i$. Multiplication and division also follow the rules for irrational numbers.

EXAMPLE

Simplify $\left(8 + 7i\right)\left(5 - 3i\right)$.

Use the distributive property (or FOIL) to get $40 + 35i - 24i - 21i^2$. The big difference here is that $i^2 = -1$ so that substitution can be made: $40 + 11i + 21 = 61 + 11i$. (Again, you can check this on your graphing calculator.)

EXAMPLE

Simplify $(5 - 3iz)(8 - 2iz)$.

$(5 - 3iz)(8 - 2iz) = 40 - 24iz - 10iz + 6i^2z^2 = 40 - 34iz - 6z^2 = 40 - 6z^2 - 43iz$.

The numbers $a + bi$ and $a - bi$ are called **complex conjugates**. If you add the two together the result is a real number, and when you subtract them the result is an imaginary number.

> The product of the two is also real and equals $a^2 + b^2$. As we did when dividing irrational numbers, we take advantage of the conjugates to ensure that the denominator of the fraction is a real number.

EXAMPLE

Simplify $\dfrac{7 + 3i\sqrt{2}}{5 - 4i\sqrt{2}}$.

Multiply numerator and denominator by the conjugate of the denominator.

$$\frac{7 + 3i\sqrt{2}}{5 - 4i\sqrt{2}} = \left(\frac{7 + 3i\sqrt{2}}{5 - 4i\sqrt{2}}\right)\left(\frac{5 + 4i\sqrt{2}}{5 + 4i\sqrt{2}}\right)$$

$$= \frac{35 + 15i\sqrt{2} + 28i\sqrt{2} + 12i^2\left(\sqrt{2}\right)^2}{25 + 16\left(\sqrt{2}\right)^2}$$

$$= \frac{35 + 43i\sqrt{2} - 24}{25 + 32} = \frac{11 + 43i\sqrt{2}}{57} = \frac{11}{57} + \frac{43i}{57}\sqrt{2}$$

EXAMPLE

Simplify $\dfrac{1 + i}{4 + ai}$.

$$\left(\frac{1 + i}{4 + ai}\right)\left(\frac{4 - ai}{4 - ai}\right) = \frac{4 + 4i - ai - ai^2}{16 + a^2} = \frac{4 + a}{16 + a^2} + \frac{4 - a}{16 + a^2}i$$

EXERCISE 4.3

Simplify each of the following.

1. $\left(a - \sqrt{-8b}\right) + \left(7a + \sqrt{-32b}\right)$
2. $\left(a - \sqrt{-8b}\right)\left(7a + \sqrt{-32b}\right)$

3. $\dfrac{a - \sqrt{-8b}}{7a + \sqrt{-32b}}$

4. $\dfrac{4 - 8i\sqrt{3}}{4 + 8i\sqrt{3}}$

The Discriminant and the Nature of the Roots of a Quadratic Equation

Computing the discriminant when solving quadratic equations can help save time because you'll know whether the quadratic can be factored.

The quadratic formula can be used to solve all quadratic equations. The expression within the radical, $b^2 - 4ac$, is called the **discriminant** because it distinguishes the type of solutions that the equation will have.

The table below shows the relationship between the value of the discriminant, the nature of the roots, and a picture of what the graph might look like. (Please note: all graphs are drawn opening up. If the leading coefficient is negative, the graph will be reflected over the x-axis.)

$b^2 - 4ac$	Roots	Graph
equals 0	roots are real, rational, and equal (double roots)	
greater than 0 and a perfect square	roots are real, rational, and unequal	
greater than 0 and not a perfect square	roots are real, irrational, and unequal	
less than 0	roots are complex conjugates	

EXAMPLE

▶ Determine the nature of the roots of the equation $4x^2 - 9x + 2 = 0$

▶ The discriminant equals $(-9)^2 - 4(4)(2) = 49$. Therefore, the roots are real, rational, and unequal.

EXAMPLE

▶ For what values of c will the roots of the equation $4x^2 - 3x + c = 0$ be complex?

▶ The discriminant for complex roots is a negative number.
Solve $(-3)^2 - 4(4)(c) < 0$.

$$9 - 16c < 0 \text{ yields } c > \frac{9}{16}.$$

EXAMPLE

▶ For what values of b will the roots of $8x^2 - bx + 2 = 0$ be double roots?

▶ The roots of a quadratic equation will be real whenever the discriminant is zero. Solve:

$$b^2 - 4(8)(2) = 0. \ b^2 - 64 = 0 \text{ gives } b = \pm 8.$$

EXERCISE 4.4

Answer these questions about the nature of the roots of a quadratic equation.

1. Determine the nature of the roots of the equation $9x^2 - 7x - 3 = 0$.

2. Determine the nature of the roots of the equation $9x^2 - 24x + 16 = 0$.

3. Determine the nature of the roots of the equation $9x^2 - 12x + 16 = 0$.

4. For what values of a will the roots of the equation $ax^2 - 8x + 5 = 0$ be real and unequal?

5. For what values of b will the roots of the equation $5x^2 + bx + b = 0$ be double roots?

Sum and Product of the Roots of a Quadratic Equation

The expansion of the expression $(px - r_1)(qx - r_2)$ is $pqx^2 - (pr_2 + qr_1)x + r_1r_2 = ax^2 + bx + c$ (as general quadratic equations are usually written). The solution to the equation $pqx^2 - (pr_2 + qr_1)x + r_1r_2 = 0$ or $(px - r_1)(qx - r_2) = 0$ is $x = \dfrac{r_1}{p}, \dfrac{r_2}{q}$. The product of the roots, $\dfrac{r_1r_2}{pq}$, is equal to $\dfrac{c}{a}$, while the sum of the roots to

$(px - r_1)(qx - r_2) = 0$ is $\dfrac{r_1}{p} + \dfrac{r_2}{q} = \dfrac{qr_1 + pr_2}{pq}$ and this is equal to $\dfrac{-b}{a}$.

The sum of the roots of the equation $ax^2 + bx + c = 0$ is $\dfrac{-b}{a}$ and the product of the roots is $\dfrac{c}{a}$.

EXAMPLE

▶ Example: Find the product of the roots to the equation $-8x^2 + 7x + 12 = 0$.

▶ Solution: The product of the roots is $\dfrac{c}{a} = \dfrac{12}{-8} = \dfrac{-3}{2}$.

EXAMPLE

▶ Find the sum of the roots to the equation $-8x^2 + 7x + 12 = 0$.

▶ The sum of the roots is $\dfrac{-b}{a} = \dfrac{-7}{-8} = \dfrac{7}{8}$.

EXAMPLE

▶ Determine the quadratic equation with integral coefficients whose roots are $\dfrac{-2}{3}$ and $\dfrac{3}{4}$.

▶ The sum of the roots $\dfrac{-2}{3} + \dfrac{3}{4} = \dfrac{1}{12} = \dfrac{-b}{a}$ while the product of the roots $\left(\dfrac{-2}{3}\right)\left(\dfrac{3}{4}\right) = \dfrac{-6}{12} = \dfrac{c}{a}$. In this case it is best not to reduce the fraction representing the product. The reason for this is that both denominators represent the leading coefficient, a, so $a = 12$. The numerator of the sum allows us to determine that $b = -1$, and the numerator of the product tells us that $c = -6$. Therefore, the equation is $12x^2 - x - 6 = 0$. (This might seem to be a trivial point, but the directions to the problem said to write a quadratic equation. For that reason, the response $12x^2 - x - 6$ would be incorrect because this is a quadratic expression, not an equation.)

EXAMPLE

Determine the quadratic equation with integral coefficients whose roots sum to $\dfrac{-7}{24}$ and whose product is $\dfrac{5}{12}$.

We have $\dfrac{-b}{a} = \dfrac{-7}{24}$ and $\dfrac{c}{a} = \dfrac{5}{12}$. In order to be able to read the coefficients, the denominators must be the same. Therefore, rewrite the product of the roots with the equivalent fraction $\dfrac{10}{24}$. You can now state that $a = 24$, $b = 7$, and $c = 10$. The correct equation is $24x^2 + 7x + 10 = 0$.

EXAMPLE

Determine the quadratic equation with integral coefficients whose roots are $\dfrac{-2}{3} + \dfrac{5}{8}i$ and $\dfrac{-2}{3} - \dfrac{5}{8}i$.

The sum of the roots is $\dfrac{-b}{a} = \dfrac{-4}{3}$ and the product is $\dfrac{c}{a} = \dfrac{4}{9} + \dfrac{25}{64} = \dfrac{481}{576}$.

The sum of the roots needs to be rewritten so that the denominator is 576.

$\dfrac{-b}{a} = \dfrac{-4}{3} = \dfrac{-768}{576}$. Therefore, the correct equation is $576x^2 + 768x + 481 = 0$.

EXERCISE 4.5

Answer the following questions.

1. Find the sum of the roots to the equation $10x^2 + 25x - 17 = 0$.

2. Find the product of the roots to the equation $10x^2 + 25x - 17 = 0$.

For problems 3–5, write a quadratic equation with integral coefficients that have the given roots.

3. $\dfrac{3}{5}, \dfrac{-7}{10}$

4. $5 + 2\sqrt{3}, 5 - 2\sqrt{3}$

5. $\dfrac{5}{8} + \dfrac{7\sqrt{3}}{12}i, \dfrac{5}{8} - \dfrac{7\sqrt{3}}{12}i$

CHAPTER 5

Polynomial Functions

Most of the work you have done with polynomials at this point has been with quadratics. You may have done some work with cubics that have fit the sum or difference of cubes formulas, with cubics that had a common variable factor, or with cubics that could be split into two pairs of terms that had a common factor. In this chapter we will look at polynomial functions of higher degree (the largest exponent in the problem) and do some analysis.

Even and Odd Functions

Some functions, such as $3x^6 + 4x^4 - 5x^2$, $3x^6 + 4x^4 - 5x^2 + 6$, and $3x^8 - 4x^6 + 7x^2$ consist of terms that have only even exponents. In the case of $3x^6 + 4x^4 - 5x^2 + 6$, the constant can be thought of as being the coefficient of x^0. Cleverly enough, because all the exponents are even, the polynomial is said to be even. The important aspect of even functions is that they are always symmetric to the y-axis. This means that $f(-x) = f(x)$ for all values of x in the domain.

Odd functions (e.g., $9x^5 + 7x^3 - 3x$) contain only odd exponents. These functions are always symmetric about the origin. This means that $f(-x) = -f(x)$ for all values of x in the domain.

Functions that contain both even and odd exponents have no special name and do not have either of the symmetries mentioned in the previous paragraphs.

EXAMPLE

▶ If $f(x)$ is an odd function, what is the value of $f(3) + f(-3)$?

▶ Since $f(-3) = -f(3)$ the sum of $f(3) + f(-3) = 0$.

EXERCISE 5.1

Classify each of the following as even, odd, or neither.

1.

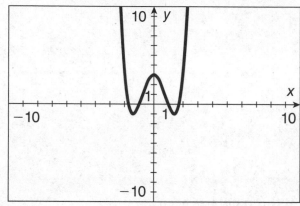

3.

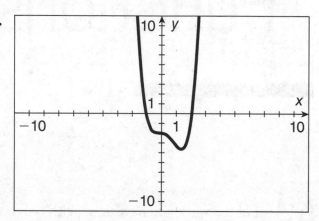

2.

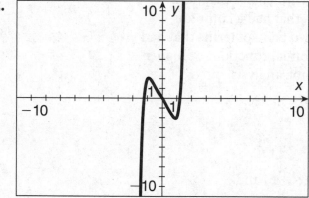

4.

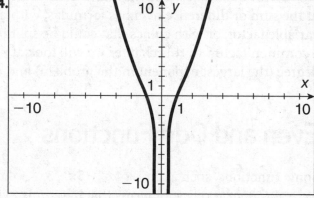

End Behavior

A topic that will come to play in future studies is the concept of end behavior. That is, what can be said about a function as x approaches negative infinity ($-\infty$) or positive infinity (∞). It turns out that no matter how many terms are involved in the polynomial, only the term of highest degree (with the biggest exponent) needs to be examined. The coefficient has great importance in this matter.

Consider the function $f(x) = 4x^7$. As x approaches $-\infty$ (written $x \to -\infty$), x^7 will be a very large negative number that when multiplied by 4 is still a large negative number. Therefore, as $x \to -\infty \Rightarrow f(x) \to -\infty$. We also know as $x \to \infty \Rightarrow f(x) \to \infty$. What does this tell us? When we look at the graph of this function, the left side of the graph will go to the bottom left corner (as x gets large in the negative direction, then y also gets large in the negative direction) while the right side of the graph will go to the upper right.

Do you see that the graph of $f(x) = -4x^7$ will reverse end behavior? That is, as $x \to -\infty \Rightarrow f(x) \to \infty$ and $x \to \infty \Rightarrow f(x) \to -\infty$.

With functions that have an even degree, the end behavior will be that $f(x)$ will go to infinity if the leading coefficient is positive or $f(x)$ will go to negative infinity if the leading coefficient is negative.

EXERCISE 5.2

Determine the end behavior for each function.

1. $f(x) = -7x^4 - 3x^3 + 8$

2. $f(x) = 6x^7 + 13x^4 + 1{,}000$

3. $f(x) = 7x^2 + 3x^9 + 8x^4 - 10$

4. $f(x) = 5x^6 - 31x + 8$

Remainder and Factor Theorems

Think back to your early elementary school days when you were first learning how to divide numbers. You learned that $12 \div 3 = 4$ because $3 \times 4 = 12$. You then were taught the language of factors: 3 and 4 are factors of 12 because when they are multiplied the product is 12. The next thing you learned is that 3 is not a factor of 14 because there is no whole number you can multiply with 3 to get 14. Along with this you learned that $14 \div 3 = 4$ remainder 2. Once you had fractions figured out, the answer to the problem $14 \div 3$ was $4\frac{2}{3}$.

Let's extend this knowledge to polynomials. If the polynomial $P(x)$ is divided by the monomial $x - a$, you will get a quotient, $Q(x)$, and a remainder, R. That is, $\dfrac{P(x)}{x-a} = Q(x) + \dfrac{R}{x-a}$. If both sides of the equation are multiplied by $x - a$, we get $P(x) = Q(x)(x - a) + R$. Finally, we get to the big idea at hand. If we evaluate $P(x)$ with $x = a$, we get $P(a) = Q(a)(0) + R = R$. We call this the **Remainder Theorem**. An immediate consequence of this is the **Factor Theorem**: If $P(a) = 0$ then $x - a$ is a factor of $P(x)$.

▶ What is the remainder when $4x^3 - 5x^2 + 12x - 8$ is divided by $x - 3$?

▶ Evaluate $4x^3 - 5x^2 + 12x - 8$ with $x = 3$ to get $4(3)^3 - 5(3)^2 + 12(3) - 8 = 91$.

▶ What is the remainder when $4x^3 - 5x^2 + 12x - 8$ is divided by $x + 4$?

▶ Replace x with -4 (because the form is supposed to be $x - a$, so $x + 4$ is written as $x - (-4)$).

$$4(-4)^3 - 5(-4)^2 + 12(-4) - 8 = -392.$$

▶ What is the remainder when $4x^3 - 5x^2 + 12x - 8$ is divided by $2x + 1$?

▶ Rewrite $2x + 1$ as $2\left(x - \dfrac{-1}{2}\right)$ to see that we need to evaluate the polynomial with $\dfrac{-1}{2}$.

$$4\left(\frac{-1}{2}\right)^3 - 5\left(\frac{-1}{2}\right)^2 + 12\left(\frac{-1}{2}\right) - 8 = -15.75.$$

▶ Is $x - 3$ a factor of $6x^3 + x^2 - 47x - 30$?

▶ Evaluate $6x^3 + x^2 - 47x - 30$ with $x = 3$. $6(3)^3 + (3)^2 - 47(3) - 30 = 0$. Yes, $x - 3$ is a factor of $6x^3 + x^2 - 47x - 30$.

EXERCISE 5.3

Use the Factor Theorem to determine if the given monomial is a factor of the given polynomial, P(x).

1. $x + 3$; $P(x) = 16x^3 + 48x^2 - x - 3$

2. $x - 5$; $P(x) = 6x^3 - 37x^2 + 32x + 15$

3. $2x - 3$; $P(x) = 6x^3 - 37x^2 + 32x + 15$

4. $x + 6$; $P(x) = 4x^4 + 4x^3 - 129x^2 - 9x + 270$

5. $x - 6$; $P(x) = 4x^4 + 4x^3 - 129x^2 - 9x + 270$

6. $x - 4$; $P(x) = 9x^4 + 18x^3 - 97x^2 - 50x + 200$

7. $3x - 5$; $P(x) = 9x^4 + 18x^3 - 97x^2 - 50x + 200$

8. $2x - 5$; $P(x) = 2x^4 + 11x^3 + 4x^2 - 62x - 120$

Synthetic Division

Let's go back to elementary school again. You've learned how to divide. Now you are asked to learn how to factor larger numbers into their prime factors (primarily designed to help you with finding common denominators for adding and subtracting fractions).

As with most of algebra, which variable is chosen for the problem has no impact on the solution to the problem. Synthetic division eliminates the variables and allows you to concentrate on the variables that dictate the quotient and remainder.

EXAMPLE

▶ Find the prime factors of 72.

▶ First, factor 72 to 2 and 36.

▶ Then factor 36 to 2 and 18.

▶ Factor 18 to 2 and 9.

▶ Factor 9 to 3 and 3.

▶ All the factors are prime numbers.
Therefore, $72 = 2^3 \times 3^2$.

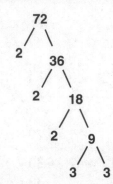

We can use a similar process to determine the prime factors of polynomials. Once we find a factor, we can divide the polynomial by it to get another factor and then continue to reduce the resulting polynomial until it no longer factors. We saw in the last section that $x - 3$ is a factor of $6x^3 + x^2 - 47x - 30$. We can divide $6x^3 + x^2 - 47x - 30$ by $x - 3$ as we did in Algebra I (remember how easy it is to make a mistake with subtraction). Instead we will use a process called synthetic division. There are a few differences from the long division we used to do. First, we ignore the variables and just work with the coefficients. Second, as we did with the Remainder Theorem, rather than work with $x - a$, we just work with a. Third, rather than subtract, we add.

Write the coefficients for the terms in order:

$$6 \quad 1 \quad -47 \quad -30$$

Write the constant from the divisor to the left of the coefficients:

$$3| \quad 6 \quad 1 \quad -47 \quad -30$$

Bring the first coefficient "down":

$$
\begin{array}{r|rrrr}
3 & 6 & 1 & -47 & -30 \\
  & \\ \hline
  & 6
\end{array}
$$

Multiply the constant and the leading coefficient. Write the product under the second coefficient.

$$
\begin{array}{r|rrrr}
3 & 6 & 1 & -47 & -30 \\
  &   & 18 \\ \hline
  & 6
\end{array}
$$

Add the numbers in the second column. Write the sum below the line.

$$
\begin{array}{r|rrrr}
3 & 6 & 1 & -47 & -30 \\
  &   & 18 &   &   \\
\hline
  & 6 & 19 &   &   \\
\end{array}
$$

Multiply the constant with this result. Write the product in the next column.

$$
\begin{array}{r|rrrr}
3 & 6 & 1 & -47 & -30 \\
  &   & 18 & 57 &   \\
\hline
  & 6 & 19 &   &   \\
\end{array}
$$

Add:

$$
\begin{array}{r|rrrr}
3 & 6 & 1 & -47 & -30 \\
  &   & 18 & 57 &   \\
\hline
  & 6 & 19 & 10 &   \\
\end{array}
$$

Multiply:

$$
\begin{array}{r|rrrr}
3 & 6 & 1 & -47 & -30 \\
  &   & 18 & 57 & 30 \\
\hline
  & 6 & 19 & 10 &   \\
\end{array}
$$

Add:

$$
\begin{array}{r|rrrr}
3 & 6 & 1 & -47 & -30 \\
  &   & 18 & 57 & 30 \\
\hline
  & 6 & 19 & 10 & 0 \\
\end{array}
$$

The last entry is the remainder. Since $x^3 \div x = x^2$, we know that the quotient is a quadratic expression. Therefore, we have $6x^3 + x^2 - 47x - 30) \div (x - 3) = 6x^2 + 19x + 10$. You can show that $6x^2 + 19x + 10 = (3x + 2)(2x + 5)$, so now you have $6x^3 + x^2 - 47x - 30 = (x - 3)(3x + 2)(2x + 5)$.

▶ Divide $4x^3 - 5x^2 + 12x - 8$ by $x + 4$.

▶ We use -4 as the constant because $x + 4$ is written as $x - (-4)$.

▶ Set up the constant and the coefficients.

$$
\begin{array}{r|rrrr}
-4 & 4 & -5 & 12 & -8 \\
   &   &   &   &   \\
\hline
   &   &   &   &   \\
\end{array}
$$

▶ Bring down the 4:

$$
\begin{array}{r|rrrr}
-4 & 4 & -5 & 12 & -8 \\
   &   &   &   &   \\
\hline
   & 4 &   &   &   \\
\end{array}
$$

▶ Multiply and add:

$$
\begin{array}{r|rrrr}
-4 & 4 & -5 & 12 & -8 \\
   &   & -16 &   &   \\
\hline
   & 4 & -21 &   &
\end{array}
$$

▶ Multiply and add:

$$
\begin{array}{r|rrrr}
-4 & 4 & -5 & 12 & -8 \\
   &   & -16 & 84 &   \\
\hline
   & 4 & -21 & 96 &
\end{array}
$$

▶ Multiply and add:

$$
\begin{array}{r|rrrr}
-4 & 4 & -5 & 12 & -8 \\
   &   & -16 & 84 & -384 \\
\hline
   & 4 & -21 & 96 & -392
\end{array}
$$

▶ Therefore, $4x^3 - 5x^2 + 12x - 8$ divided by $x + 4$ is $4x^2 - 21x + 96 + \dfrac{-392}{x+4}$.

You can look back to the example in the last section where we determined that the remainder was –392.

EXAMPLE

▶ Divide $9x^4 + 63x^3 + 50x^2 - 28x - 24$ by $x + 6$.

▶ Set up:

$$
\begin{array}{r|rrrrr}
-6 & 9 & 63 & 50 & -28 & -24 \\
\hline
   &   &   &   &   &
\end{array}
$$

▶ Bring down the leading coefficient:

$$
\begin{array}{r|rrrrr}
-6 & 9 & 63 & 50 & -28 & -24 \\
\hline
   & 9 &   &   &   &
\end{array}
$$

▶ Multiply and add:

$$
\begin{array}{r|rrrrr}
-6 & 9 & 63 & 50 & -28 & -24 \\
   &   & -54 &   &   &   \\
\hline
   & 9 & 9 &   &   &
\end{array}
$$

▶ Multiply and add:

$$
\begin{array}{r|rrrrr}
-6 & 9 & 63 & 50 & -28 & -24 \\
   &   & -54 & -54 &   &   \\
\hline
   & 9 & 9 & -4 &   &
\end{array}
$$

▶ Multiply and add:

$$
\begin{array}{r|rrrrr}
-6| & 9 & 63 & 50 & -28 & -24 \\
 & & -54 & -54 & 24 & \\
\hline
 & 9 & 9 & -4 & -4 &
\end{array}
$$

▶ Multiply and add:

$$
\begin{array}{r|rrrrr}
-6| & 9 & 63 & 50 & -28 & -24 \\
 & & -54 & -54 & 24 & 24 \\
\hline
 & 9 & 9 & -4 & -4 & 0
\end{array}
$$

▶ Therefore, $9x^4 + 63x^3 + 50x^2 - 28x - 24$ divided by $x + 6$ is $9x^3 + 9x^2 - 4x - 4$. The question now becomes, how do we factor $9x^3 + 9x^2 - 4x - 4$? This is an "easier" problem because the coefficients repeat the way they do. We know that substituting 1 for x will give 14. Do you see that using -1 for x will give 0? (This is why the problem is "easier"—the coefficients will cancel each other out.)

▶ Set up:

$$
\begin{array}{r|rrrr}
-1| & 9 & 9 & -4 & -4 \\
\hline
 & & & &
\end{array}
$$

▶ Bring down the leading coefficient:

$$
\begin{array}{r|rrrr}
-1| & 9 & 9 & -4 & -4 \\
\hline
 & 9 & & &
\end{array}
$$

▶ Multiply and add:

$$
\begin{array}{r|rrrr}
-1| & 9 & 9 & -4 & -4 \\
 & & -9 & & \\
\hline
 & 9 & 0 & &
\end{array}
$$

▶ Multiply and add:

$$
\begin{array}{r|rrrr}
-1| & 9 & 9 & -4 & -4 \\
 & & -9 & 0 & \\
\hline
 & 9 & 0 & -4 &
\end{array}
$$

▶ Multiply and add:

$$
\begin{array}{r|rrrr}
-1| & 9 & 9 & -4 & -4 \\
 & & -9 & 0 & 4 \\
\hline
 & 9 & 0 & -4 & 0
\end{array}
$$

▶ $9x^4 + 63x^3 + 50x^2 - 28x - 24$ factors to:

$$(x+6)(x+1)(9x^2-4) = (x+6)(x+1)(3x-2)(3x+2).$$

EXAMPLE

This gives us the chance to talk about dividing a polynomial by a term of the form $ax + b$. Divide $9x^4 + 63x^3 + 50x^2 - 28x - 24$ by $3x - 2$. We know how to set up the coefficients of the polynomial. When we solve $3x - 2 = 0$, we get $x = \dfrac{2}{3}$.

▶ Set up:

$$
\dfrac{2}{3}\,\bigg|\quad 9 \qquad 63 \qquad 50 \qquad -28 \qquad -24
$$

▶ Bring down the leading coefficient, and continue the process:

$$
\dfrac{2}{3}\,\bigg|\quad
\begin{array}{rrrrr}
9 & 63 & 50 & -28 & -24 \\
 & 6 & 46 & 64 & 24 \\
\hline
9 & 69 & 96 & 36 & 0
\end{array}
$$

▶ We have $\left(9x^4 + 63x^3 + 50x^2 - 28x - 24\right) \div 3\left(x - \dfrac{2}{3}\right) =$

$\left(9x^3 + 69x^2 + 96x + 36\right) \div 3 = 3x^3 + 23x^2 + 32x + 12$.

> In synthetic division, when dividing by a term of the form $ax + b$, first divide by $x + \dfrac{b}{a}$ and then divide by a.

In an earlier chapter we showed how to use a graph to determine the roots, and then from the roots we can determine the factors. The same is true here.

EXAMPLE

▶ Sketch the graph of $f(x) = 9x^4 + 63x^3 + 50x^2 - 28x - 24$.

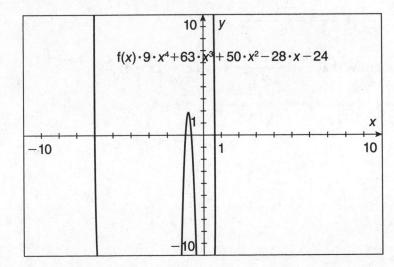

▶ Use the Zero or Intercept feature on your graphing calculator to get the zeroes of the function, –6, –1.5, –1, and 1.5 and convert these to the roots of the equation.

EXAMPLE

Divide $4x^3 - 10x^2 + 8$ by $x - 2$.

The thing that is different in this problem is that we have terms with x^3 and x^2 but no term in x. We accommodate this by using 0 as a coefficient.

Set up:

$$2|\quad 4 \quad -10 \quad 0 \quad 8$$

Bring down the first term:

$$2|\quad 4 \quad -10 \quad 0 \quad 8$$
$$\overline{4}$$

Multiply and add:

$$2|\quad 4 \quad -10 \quad 0 \quad 8$$
$$8$$
$$\overline{4 \quad -2}$$

Multiply and add:

$$2|\quad 4 \quad -10 \quad 0 \quad 8$$
$$8 \quad -4$$
$$\overline{4 \quad -2 \quad -4}$$

Multiply and add:

$$2|\quad 4 \quad -10 \quad 0 \quad 8$$
$$8 \quad -4 \quad -8$$
$$\overline{4 \quad -2 \quad -4 \quad 0}$$

We now know that $\dfrac{4x^3 - 10x^2 + 8}{x + 3}$ is $4x^2 - 2x - 4$.

EXERCISE 5.4

Use synthetic division to find the quotient and remainder (if any) for each of the following.

1. $(16x^3 + 48x^2 - x - 3) \div (x + 3)$

2. $(6x^3 - 37x^2 + 32x + 15) \div (x - 5)$

3. $(6x^3 - 37x^2 + 32x + 15) \div (2x - 3)$

4. $(4x^4 + 4x^3 - 129x^2 - 9x + 270) \div (x + 6)$

5. $(4x^4 + 4x^3 - 129x^2 - 9x + 270) \div (x - 6)$

6. $(9x^4 + 18x^3 - 97x^2 - 50x + 200) \div (x - 4)$

7. $(9x^4 + 18x^3 - 97x^2 - 50x + 200) \div (3x - 5)$

8. $(2x^4 + 11x^3 + 4x^2 - 62x - 120) \div (2x - 5)$

9. $(x^5 + 4x^4 - 3x^2 + 5x - 10) \div (x + 2)$

Factor completely:

10. $16x^3 + 48x^2 - x - 3$

11. $6x^3 - 37x^2 + 32x + 15$

12. $9x^4 + 18x^3 - 97x^2 - 50x + 200$

13. $4x^4 + 4x^3 - 129x^2 - 9x + 270$

14. $x^3 - 5x - 12$

Fundamental Theorem of Algebra

As part of his doctoral thesis, Carl Gauss proved that for every polynomial equation of degree n, there are n roots from the set of complex numbers. This statement is called the **Fundamental Theorem of Algebra**. With the exception of the work we did with complex numbers in Chapter 4, we have only considered solutions from the set of real numbers.

EXAMPLE

▶ Solve $x^3 - 1 = 0$.

▶ According to the Fundamental Theorem of Algebra, there are three solutions to this problem. Obviously, $x = 1$ is one of the solutions. To find the other roots, factor the cubic.

$$x^3 - 1 = (x - 1)(x^2 + x + 1).$$

▶ Set each term equal to 0 to get:

$x - 1 = 0$ so $x = 1$ or $x^2 + x + 1 = 0$ so:

$$x = \frac{-1 \pm \sqrt{(1)^2 - 4(1)(1)}}{2} = \frac{-1}{2} \pm \frac{\sqrt{3}}{2} i.$$

EXAMPLE

Solve $2x^4 + 11x^3 + 4x^2 - 62x - 120 = 0$

A portion of the graph of $y = 2x^4 + 11x^3 + 4x^2 - 62x - 120$ is shown below.

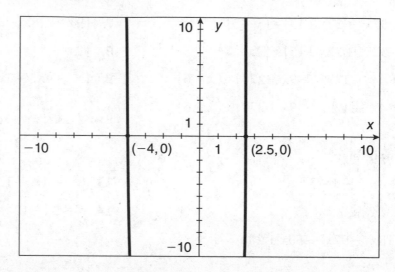

Using the Zero feature on the calculator, we find that the graph crosses the x-axis when $x = -4$ and when $x = 2.5$.

This tells us that $x + 4$ and $2x - 5$ are factors of $2x^4 + 11x^3 + 4x^2 - 62x - 120$. Since there are no other x-intercepts on this graph, the remaining roots must be complex numbers. Use synthetic division to find the remaining factor.

Set up:

$$-4 \,|\quad 2 \quad 11 \quad 4 \quad -62 \quad -120$$

Bring down the leading coefficient:

$$-4 \,|\quad 2 \quad 11 \quad 4 \quad -62 \quad -120$$
$$\overline{\quad\quad 2}$$

Multiply and add:

$$-4 \,|\quad 2 \quad 11 \quad 4 \quad -62 \quad -120$$
$$\quad\quad\quad -8$$
$$\overline{\quad\quad 2 \quad 3}$$

Multiply and add:

$$-4 \,|\quad 2 \quad 11 \quad 4 \quad -62 \quad -120$$
$$\quad\quad\quad -8 \quad -12$$
$$\overline{\quad\quad 2 \quad 3 \quad -8}$$

▶ Multiply and add:

$$
\begin{array}{r|rrrrr}
-4| & 2 & 11 & 4 & -62 & -120 \\
 & & -8 & -12 & 32 & \\
\hline
 & 2 & 3 & -8 & -30 &
\end{array}
$$

▶ Multiply and add:

$$
\begin{array}{r|rrrrr}
-4| & 2 & 11 & 4 & -62 & -120 \\
 & & -8 & -12 & 32 & 120 \\
\hline
 & 2 & 3 & -8 & -30 & 0
\end{array}
$$

▶ Now use synthetic division on the reduced polynomial with $x = \dfrac{5}{2}$.

▶ Set up:

$$
\begin{array}{r|rrrr}
\dfrac{5}{2}| & 2 & 3 & -8 & -30 \\
\hline
 & & & &
\end{array}
$$

▶ Bring down the lead coefficients:

$$
\begin{array}{r|rrrr}
\dfrac{5}{2}| & 2 & 3 & -8 & -30 \\
\hline
 & 2 & & &
\end{array}
$$

▶ Multiply and add:

$$
\begin{array}{r|rrrr}
\dfrac{5}{2}| & 2 & 3 & -8 & -30 \\
 & & 5 & & \\
\hline
 & 2 & 8 & &
\end{array}
$$

▶ Multiply and add:

$$
\begin{array}{r|rrrr}
\dfrac{5}{2}| & 2 & 3 & -8 & -30 \\
 & & 5 & 20 & \\
\hline
 & 2 & 8 & 12 &
\end{array}
$$

▶ Multiply and add:

$$
\begin{array}{r|rrrr}
\dfrac{5}{2}| & 2 & 3 & -8 & -30 \\
 & & 5 & 20 & 30 \\
\hline
 & 2 & 8 & 12 & 0
\end{array}
$$

▶ So now we have $2x^4 + 11x^3 + 4x^2 - 62x - 120 =$
$\left(x + 4\right)\left(x - \dfrac{5}{2}\right)\left(2x^2 + 8x + 12\right)$. The trinomial has a common factor of 2. We'll remove the 2 from the trinomial and multiply the 2 through the factor with the fractional constant. $\left(x + 4\right)\left(x - \dfrac{5}{2}\right)\left(2x^2 + 8x + 12\right)$ becomes $\left(x + 4\right)\left(2x - 5\right)\left(x^2 + 4x + 6\right)$. Use the quadratic formula to solve

$$
x^2 + 4x + 6 = 0, \quad x = \frac{-4 \pm \sqrt{16 - 4(1)(6)}}{2} = \frac{-4 \pm \sqrt{-8}}{2} = -2 \pm i\sqrt{2}.
$$

Solve: $x^6 - 19x^3 - 216 = 0$

This is a special type of problem called quadratic form. This is because there are three terms, the last term is a constant, and the square of the variable in the middle term appears in the first term. Sketch the graph of $y = p^2 - 19p - 216$.

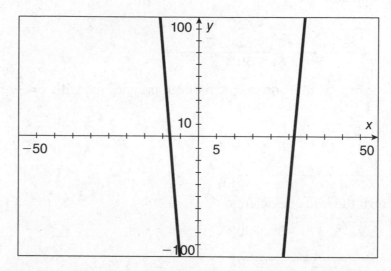

Use the Zero feature to determine that the zeros are $x = -8$ and $x = 27$. Therefore, $x^6 - 19x^3 - 216 = (x^3 + 8)(x^3 - 27) = (x + 2)(x^2 - 2x + 4)(x - 3)(x^2 + 3x + 9)$. Use the quadratic formula to solve the trinomial equations to find the solutions to $x^6 - 19x^3 - 216 = 0$ are $x = -2, 3, 1 \pm i\sqrt{3}, \dfrac{-3}{2} \pm \dfrac{3\sqrt{3}}{2}i$.

EXERCISE 5.5

Solve each equation over the set of complex numbers.

1. $4x^5 + 4x^4 - 17x^3 - 41x^2 + 18x + 72 = 0$

2. $6x^4 - 13x^3 + 30x^2 - 52x + 24 = 0$

3. $48x^4 + 164x^3 + 208x^2 - 21x - 245 = 0$

4. $48x^4 - 172x^3 + 212x^2 - 60x - 252 = 0$

5. $x^6 - 25x^4 - 16x^2 + 400 = 0$

Polynomial Inequalities

An efficient way to solve polynomial inequalities is to first solve the corresponding equations and then perform a signs analysis to the intervals formed. Since we solve inequalities over the set of real numbers, we do not need to consider the complex solutions.

> A real number must have one of three conditions: it is positive, it is negative, or it is 0. By determining when a polynomial or rational expression is equal to zero, you can determine when the expression is positive or negative.

EXAMPLE

Solve $2x^4 + 11x^3 + 4x^2 - 62x - 120 \geq 0$

We saw in the last section that the graph of $y = 2x^4 + 11x^3 + 4x^2 - 62x - 120$ is:

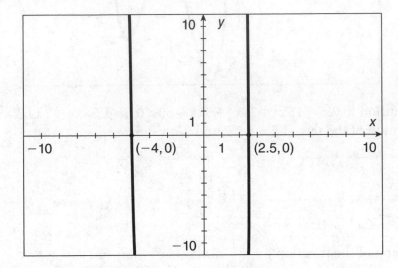

We are looking for those values of x for which the y-coordinate is either zero or positive. We can see from the graph that this happens when $x \leq -4$ or $x \geq 2.5$.

EXAMPLE

Solve $2x^4 - 15x^3 + 21x^2 + 32x - 60 \le 0$

Examine the graph of $y = 2x^4 - 15x^3 + 21x^2 + 32x - 60$

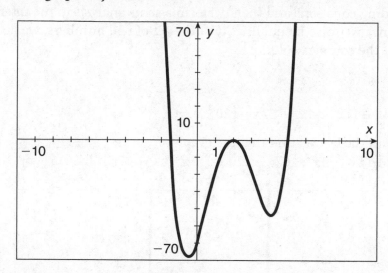

The solution to $2x^4 - 15x^3 + 21x^2 + 32x - 60 < 0$ is $-1.5 < x < 2 \cup 2 < x < 5$. (The union of the sets $-1.5 < x < 2$ and $2 < x < 5$.)

EXAMPLE

Solve $x^4 - 2x^3 - 11x^2 - 8x - 60 > 0$

The graph of the function $y = x^4 - 2x^3 - 11x^2 - 8x - 60$ is:

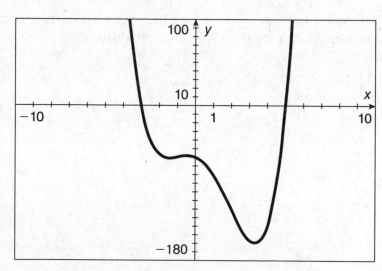

The graph lies above the x-axis when $x < -3$ or $x > 5$.

EXAMPLE

▶ For what values of b will the roots of $8x^2 - bx + 2 = 0$ be real?

▶ The roots of a quadratic equation will be real whenever the discriminant is nonnegative.

▶ Solve $b^2 - 4(8)(2) \geq 0$.

$b^2 - 64 \geq 0$ gives $b \leq 8$ or $b \geq 8$.

EXERCISE 5.6

Solve each of the given inequalities.

1. $x^3 - 9x^2 + 8x > 0$

2. $20x^3 - 87x^2 - 234x + 216 \geq 0$

3. $x^4 - 8x^3 + 26x^2 - 80x + 160 > 0$

4. $x^5 - 16x^3 - 8x^2 + 128 \leq 0$

5. $9x^6 - 70x^4 + 161x^2 - 100 < 0$

CHAPTER 6

Rational and Irrational Functions

The rules for dealing with rational expressions are exactly the same as when you first saw them in middle school:

- Don't divide by zero.
- When multiplying, look for common factors so that the terms can be simplified.
- When dividing, invert the divisor, change the operation to multiplication, and follow the rules for multiplication.
- When adding and subtracting, find a common denominator.

Rational Functions

Functions of the form $f(x) = \dfrac{g(x)}{h(x)}$ are called rational functions because they are the ratio of two functions. Being able to determine the domain, end behavior, and range of these functions are important skills.

EXAMPLE

Find the domain for each function.

(a) $p(x) = \dfrac{5x^2 + 3}{x + 1}$

(b) $q(x) = \dfrac{3x^2 - 5x - 7}{4x^2 - 5x + 1}$

(c) $z(x) = \dfrac{x^2 + 4x - 5}{x^3 - 3x^2 + 2x}$

The domain for $p(x)$ is $x \neq -1$, the domain for $q(x)$ is $x \neq 1, \frac{1}{4}$, and the domain for $z(x)$ is $x \neq 0, 1, 2$ because in each case these values of x would cause the denominator of the fraction to equal 0.

End behavior is not always as easy to determine as it is for polynomial functions. We pay attention to the degrees (largest exponent) of the polynomials that form the numerators and denominators. There are three different scenarios that can occur:

- The degree of the numerator is greater then the degree of the denominator.
- The degree of the numerator equals the degree of the denominator.
- The degree of the numerator is less than the degree of the denominator.

Let's use the functions $p(x)$, $q(x)$, and $z(x)$ above to examine end behaviors.

(a) **$p(x)$.** The degree of the numerator is greater than the degree of the denominator. In middle school, you would have called this an improper fraction and could alter the form to a mixed number. In essence, that is what can be done here. Divide $5x^2 + 3$ by $x + 1$ to get $5x - 5 + \dfrac{8}{x+1}$. As $x \to -\infty$ $5x - 5 \to -\infty$ and $\dfrac{8}{x+1} \to 0$. Therefore, we conclude as $x \to -\infty$, $p(x) \to -\infty$. As $x \to \infty$ $5x - 5 \to \infty$ and $\dfrac{8}{x+1} \to 0$. Therefore, we conclude as $x \to \infty$, $p(x) \to \infty$.

(b) **$q(x)$.** The degrees of the numerator and denominator are equal. In this case, divide both the numerator and denominator by x^2, the term of highest degree, rewriting $\dfrac{3x^2 - 5x - 7}{4x^2 - 5x + 1}$ as $\dfrac{3 - \dfrac{5}{x} - \dfrac{7}{x^2}}{4 - \dfrac{5}{x} + \dfrac{1}{x^2}}$. As $x \to \pm\infty$, the terms $\dfrac{5}{x}, \dfrac{7}{x^2}$, and $\dfrac{1}{x^2}$ all go to zero. Therefore, as $x \to \pm\infty$, $q(x) \to \dfrac{3}{4}$.

(c) **$z(x)$.** The degree of the numerator is less than the degree of the denominator. Divide all terms by the degree of the largest term, x^3, to rewrite $\dfrac{x^2 + 4x - 5}{x^3 - 3x^2 + 2x}$ as $\dfrac{\dfrac{1}{x} + \dfrac{4}{x^2} - \dfrac{5}{x^3}}{1 - \dfrac{3}{x} + \dfrac{2}{x^2}}$. All terms in the numerator go to zero as $x \to \pm\infty$ while the denominator always goes to 1. Therefore, as $x \to \pm\infty$, $z(x) \to 0$.

Based on these examples, we can deduce that:

- When the degree of the numerator is greater than the degree of the denominator, the end behavior will be that the graph will go to $\pm\infty$ depending on the sign of the coefficients and whether the degree of the polynomial is even or odd.
- When the degree of the numerator is equal to the degree of the denominator, the end behavior will be the ratio of the coefficients of the terms of highest degree.
- When the degree of the denominator is less than the degree of the numerator, the end behavior is that the expression will go to 0.

In Chapter 2, we saw that we could use the inverse of a function to determine the range of the function. While this is always a true statement, the problems that arise are that the inverse of a rational function might not exist because the function was not one-to-one or the algebra for determining the inverse might be very difficult. Given these situations, we'll often look at the graph of a function using our graphing technology to help us.

p(x)

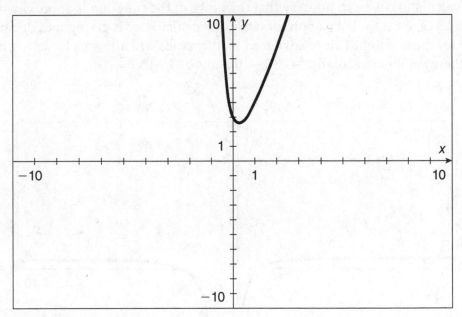

This looks like a bit like a parabola with a smallest value of 2.65 (use the minimum feature on your calculator). Do you see what the problem is? We just determined a few paragraphs above that as $x \to \infty$, $p(x) \to \infty$. There must be more to the picture. Use the Zoom–Fit feature on your calculator to get a better picture.

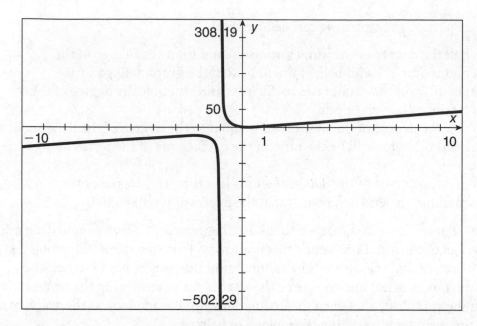

This picture gives a more accurate representation of the graph. The left branch of this curve has a maximum value of –22.9 (use the max feature on your calculator). We can now say that the range of the function is approximately $y \leq -22.9$ or $y \geq 2.65$. Why approximately? Our calculator is giving us a decimal value for those points. (More advanced mathematics will allow us to determine that these values are actually $-4\sqrt{10} - 10$ and $4\sqrt{10} - 10$.)

$q(x)$

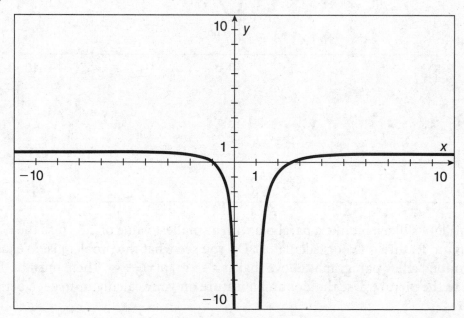

We found that the domain of this function is $x \neq 1, \frac{1}{4}$. The graph shown shows what is happening to the left of $\frac{1}{4}$ and to the right of 1. What is happening between these values? Use Zoom—Fit to see.

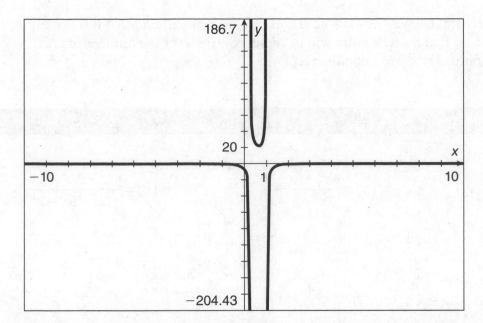

The minimum value for the upper section of this graph is 15.9. Therefore, the range of the $q(x)$ is approximately $y < 0.75$ or $y \geq 15.9$. ($y < 0.75$ rather than $\leq$ because we saw that as $x \to \pm \infty$, $q(x) \to \dfrac{3}{4}$.)

$z(x)$

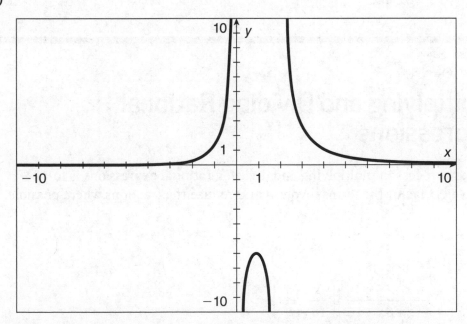

The lower piece of the graph is visible so there is no need to zoom. The maximum value of the lower region is approximately –5.96. Therefore, the range of $z(x)$ is approximately $y \leq -5.96$ or $y > 0$. Wait, there's more! Look at the factored form of the function $z(x) = \dfrac{x^2 + 4x - 5}{x^3 - 3x^2 + 2x} = \dfrac{(x+5)(x-1)}{x(x-2)(x-1)} = \dfrac{x+5}{x(x-2)}$.

The graph makes it look like there is a point on the graph where $x = 1$. There can't be because 1 is not in the domain of the function. Consequently, we need to remove the point at which $x = 1$, and that is (1, –6). Therefore, the range is approximately $y \leq -5.96$ or $y > 0$ and $y \neq -6$. You will find that it is usually the case that a single point will be removed whenever the numerator and denominator share a common factor.

EXERCISE 6.1

Given:

$$r(x) = \frac{x^2 - 9}{x + 2} \qquad t(x) = \frac{4x^2 + 3x + 2}{2x^2 + x - 1} \qquad v(x) = \frac{6x^3 + 2x^2 - 4x}{3x^2 - 5x + 2}$$

Determine the domain for:

1. $r(x)$ **2.** $t(x)$ **3.** $v(x)$

Find the end behavior for:

4. $r(x)$ **5.** $t(x)$ **6.** $v(x)$

Find the range for:

7. $r(x)$ **8.** $t(x)$ **9.** $v(x)$

> An important concept in reducing terms is that $\frac{x - a}{a - x} = -1$.

Multiplying and Dividing Rational Expressions

The key process in multiplying and dividing rational expressions is to completely factor the terms involved and reduce the fractions where possible.

EXAMPLE

Simplify:

$$\left(\frac{x^2 - 4x}{4x^2 + 8x + 16} \right)\left(\frac{x^3 - 8}{16 - x^2} \right)$$

Factor each of the terms:

$$\frac{x(x - 4)}{4(x^2 + 2x + 4)} \times \frac{(x - 2)(x^2 + 2x + 4)}{(4 - x)(4 + x)} = \frac{x(x - 4)^{-1}}{4(x^2 + 2x + 4)} \times \frac{(x - 2)(x^2 + 2x + 4)}{(4 - x)(4 + x)}$$

$$= \frac{-x(x - 2)}{4(4 + x)}$$

EXAMPLE

Simplify:

$$\left(\frac{12x^3 + 20x^2 - 75x - 125}{16x^4 - 625}\right)\left(\frac{12x^3 + 32x^2 + 75 + 200}{6x^2 + 31x + 40}\right)$$

▶ Factor each of the terms:

$$\frac{4x^2(3x+5) - 25(3x+5)}{(4x^2 - 25)(4x^2 + 25)} \times \frac{4x^2(3x+8) + 25(3x+8)}{(2x+5)(3x+8)}$$

$$\frac{(3x+5)\cancel{(4x^2 - 25)}}{\cancel{(4x^2 - 25)}\cancel{(4x^2 + 25)}} \times \frac{\cancel{(3x+8)}\cancel{(4x^2 + 25)}}{(2x+5)\cancel{(3x+8)}} = \frac{3x+5}{2x+5}$$

▶ Notice that the difference of squares was not factored in this case because it was a common factor to the numerator and the denominator.

EXAMPLE

Simplify:

$$\left(\frac{6x^3 - 37x^2 - 9x + 90}{8x^3 - 20x^2 - 18x + 45}\right)\left(\frac{8x^3 - 12x^2 - 50x + 75}{20x^3 + 80x^2 + 75x}\right)$$

▶ Factor each of the terms:

$$\frac{6x^3 - 37x^2 - 9x + 90}{4x^2(2x-5) - 9(2x-5)} \times \frac{4x^2(2x-3) - 25(2x-3)}{5x(4x^2 + 16x + 15)}$$

$4x^2(2x-5) - 9(2x-5)$ factors to

$(2x-5)(4x^2 - 9) = (2x-5)(2x-3)(2x+3)$

$4x^2(2x-3) - 25(2x-3)$ factors to

$(2x-3)(4x^2 - 25) = (2x-3)(2x-5)(2x+5)$

$5x(4x^2 + 16x + 15)$ factors to

$5x(2x+3)(2x+5)$

▶ But what about $6x^3 - 37x^2 - 9x + 90$? There are no common factors, and the expression cannot be divided into 2 pairs of terms that share a common factor. Don't forget to use your calculator to graph the corresponding polynomial function and determine the zeros of the function.

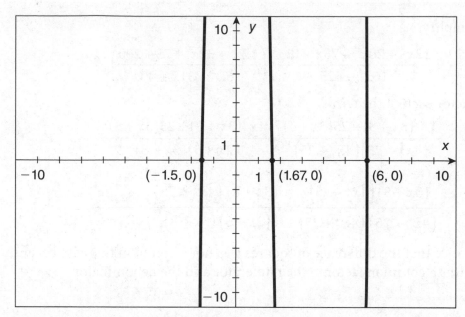

The zeros are $x = \dfrac{-3}{2}, \dfrac{5}{3}, 6$ so the factors are $(2x+3)(3x-5)(x-6)$. So we have:

$$\frac{(2x+3)(3x-5)(x-6)}{(2x-5)(2x-3)(2x+3)} \times \frac{(2x-3)(2x-5)(2x+5)}{5x(2x+3)(2x+5)} = \frac{(3x-5)(x-6)}{5x(2x+3)}$$

Division of rational expressions requires that you use the reciprocal of the divisor and change the problem to multiplication (aka, invert and multiply). Then proceed as a multiplication problem.

EXAMPLE

Simplify:

$$\frac{4x^4 - 9x^2}{4x^2 - 12x + 9} \div \frac{6x^2 + 19x + 15}{2x^3 - 7x^2 + 6x}$$

Invert and multiply:

$$\frac{4x^4 - 9x^2}{4x^2 - 12x + 9} \times \frac{2x^3 - 7x^2 + 6x}{6x^2 + 19x + 15} = \frac{x^2(4x^2 - 9)}{4x^2 - 12x + 9} \times \frac{x(2x^2 - x - 6)}{6x^2 + 19x + 15}$$

$$= \frac{x^2(2x-3)(2x+3)}{(2x-3)^2} \times \frac{x(2x-3)(x-2)}{(2x+3)(3x+5)}$$

$$= \frac{x^2(2x-3)(2x+3)}{(2x-3)^2} \times \frac{x(2x-3)(x-2)}{(2x+3)(3x+5)}$$

$$= \frac{x^3(x-2)}{3x+5}$$

EXERCISE 6.2

Simplify each of the following.

1. $\left(\dfrac{6x^2 + x - 15}{20x^2 - 9x - 20} \right)\left(\dfrac{15x^2 + 2x - 8}{9x^2 + 9x - 10} \right)$

2. $\left(\dfrac{36x^2 - 25}{54x^2 + 117x + 60} \right)\left(\dfrac{12x^2 - 26x - 56}{12x^2 - 52x + 35} \right)$

3. $\left(\dfrac{12x^2 - 53x + 56}{8x^2 - 2x - 15} \right)\left(\dfrac{4x^2 + 25x + 25}{-8x^2 - 22x + 63} \right)$

4. $\left(\dfrac{x^3 - 216}{5x^2 - 180} \right)\left(\dfrac{7x^2 + 21x - 126}{4x^2 + 24x + 144} \right)$

5. $\left(\dfrac{x^4 - 81}{x^3 - 3x^2 + 9x - 27} \right)\left(\dfrac{8x^2 - 24x + 72}{x^3 + 27} \right)$

6. $\left(\dfrac{4x^3 - 20x^2 - 9x + 45}{6x^2 - 17x - 14} \right)\left(\dfrac{12x^3 + 8x^2 + 27x + 18}{16x^4 - 81} \right)$

7. $\left(\dfrac{12x^2 + 25x + 12}{10x^2 + 29x + 10} \right) \div \left(\dfrac{35x^2 + 29x + 6}{21x^2 + 37x + 12} \right)$

8. $\left(\dfrac{8x^3 + 10x^2 - 3x}{36x^2 + 78x + 36} \right) \div \left(\dfrac{8x^2 + 22x - 6}{6x^3 + 34x^2 + 20x} \right)$

9. $\left(\dfrac{8x^3 - 22x^2 - 15x + 54}{16x^2 - 81} \right) \div \left(\dfrac{8x^3 + 27}{12x^2 + 11x - 36} \right)$

10. $\left(\dfrac{54x^3 + 33x^2 - 72x}{162x^3 + 243x^2 - 128x - 192} \right) \div \left(\dfrac{48x^4 + 390x^3 + 378x^2}{72x^2 + 145x + 72} \right)$

Adding and Subtracting Rational Expressions

The process of adding and subtracting rational expressions is consistent—first find a common denominator, combine numerators, look to see if the result can be reduced by finding common factors in the numerator and denominator. Because of this, we leave the common denominator found in factored form.

EXAMPLE

▶ Simplify:

$$\frac{x+6}{x^2-4}+\frac{x-12}{x^2+x-6}$$

▶ Factor each of the denominators:

$$\frac{x+6}{(x-2)(x+2)}+\frac{x-12}{(x+3)(x-2)}$$

▶ The term $x-2$ is common to both denominators, so the common denominator is $(x-2)(x+2)(x+3)$. Rewrite the expression with a common denominator:

$$\frac{x+6}{(x-2)(x+2)}\left(\frac{x+3}{x+3}\right)+\frac{x-12}{(x+3)(x-2)}\left(\frac{x+2}{x+2}\right)$$

▶ Combine the numerators:

$$\frac{(x+6)(x+3)+(x-12)(x+2)}{(x-2)(x+2)(x+3)}=\frac{x^2+9x+18+x^2-10x-24}{(x-2)(x+2)(x+3)}=$$

$$\frac{2x^2-x-6}{(x-2)(x+2)(x+3)}$$

▶ Factor the numerator (if possible) and reduce the fraction (if possible).

$$\frac{2x^2-x-6}{(x-2)(x+2)(x+3)}=\frac{(2x+3)(x-2)}{(x-2)(x+2)(x+3)}=\frac{2x+3}{(x+2)(x+3)}.$$

EXAMPLE

▶ Simplify:

$$\frac{x-3}{x^2+3x+2}-\frac{x-6}{x^2+4x+3}$$

▶ Factor the denominators:

$$\frac{x-3}{x^2+3x+2}-\frac{x-6}{x^2+4x+3}=\frac{x-3}{(x+1)(x+2)}-\frac{x-6}{(x+1)(x+3)}$$

▶ Get a common denominator:

$$\left(\frac{x-3}{(x+1)(x+2)}\right)\left(\frac{x+3}{x+3}\right) - \left(\frac{x-6}{(x+1)(x+3)}\right)\left(\frac{x+2}{x+2}\right) =$$

$$\frac{(x-3)(x+3)-(x-6)(x+2)}{(x+1)(x+2)(x+3)}$$

▶ Simplify the numerator making sure you remember to distribute the negative sign through the parentheses:

$$\frac{x^2-9-\left(x^2-4x-12\right)}{(x+1)(x+2)(x+3)} = \frac{x^2-9-x^2+4x+12}{(x+1)(x+2)(x+3)} = \frac{4x+3}{(x+1)(x+2)(x+3)}$$

EXAMPLE

▶ Simplify:

$$\frac{x+1}{x+2} + \frac{x-1}{x+3} - \frac{4}{(x+2)(x+3)}$$

▶ The common denominator is the product $(x+2)(x+3)$. Rewrite each of the terms with this denominator:

$$\frac{(x+1)(x+3)+(x-1)(x+2)-4}{(x+2)(x+3)}$$

▶ Simplify the numerator:

$$\frac{x^2+4x+3+x^2+x-2-4}{(x+2)(x+3)} = \frac{2x^2+5x-3}{(x+2)(x+3)} = \frac{(2x-1)(x+3)}{(x+2)(x+3)} = \frac{(2x-1)}{(x+2)}.$$

EXAMPLE

▶ Explain why $\dfrac{(x+1)(x+3)+(x-1)(x+2)-4}{(x+2)(x+3)}$ does not

become $\dfrac{(x+1)\cancel{(x+3)}+(x-1)\cancel{(x+2)}-4}{\cancel{(x+2)}\,\cancel{(x+3)}} = x+1+x-1-4 = -4$.

▶ The notion of cancellation is an act of division. Therefore, one must find a common factor in both the numerator and denominator. While $x+3$ is a factor of $(x+1)(x+3)$ in the denominator, it is not a common factor of the entire numerator and therefore cannot be removed from the problem through division.

EXERCISE 6.3

Simplify each of the following.

1. $\dfrac{2}{3(x+1)} - \dfrac{1}{2(x-1)}$

2. $\dfrac{x+1}{x+2} + \dfrac{x-3}{x-2}$

3. $\dfrac{x+1}{x^2+x-2} - \dfrac{x+2}{x^2-4}$

4. $\dfrac{2x+5}{x-4} - \dfrac{x+4}{2x-5}$

5. $\dfrac{5x+2}{2x^2-5x+2} + \dfrac{x+1}{4x^2-1}$

6. $\dfrac{x+2}{2x+3} - \dfrac{x-5}{3x-4} + \dfrac{2x^2-10x-11}{(2x+3)(3x-4)}$

Complex Fractions

A complex fraction is a fraction whose numerator and/or denominator contain fractions. For example, $\dfrac{\dfrac{1}{x}-\dfrac{1}{y}}{xy}$ is a complex fraction. A technique is to multiply the numerator and denominator of the entire fraction by the common denominator of the component fractions like so: $\dfrac{\dfrac{1}{x}-\dfrac{1}{y}}{xy}\left(\dfrac{xy}{xy}\right) = \dfrac{\dfrac{xy}{x}-\dfrac{xy}{y}}{(xy)^2} = \dfrac{y-x}{x^2y^2}$.

EXAMPLE

Simplify: $\dfrac{\dfrac{y}{x}-\dfrac{x}{y}}{x-y}$

The product xy is still the common denominator:

$$\dfrac{\dfrac{y}{x}-\dfrac{x}{y}}{x+y}\left(\dfrac{xy}{xy}\right) = \dfrac{y^2-x^2}{xy(x-y)} = \dfrac{(y-x)(y+x)}{xy(x-y)}$$

Recalling that $\dfrac{y-x}{x-y} = -1$, reduce the fraction to get:

$$\dfrac{-(y+x)}{xy}$$

Complex fractions can look overwhelming. Finding common denominators and working with smaller sections of the complex fraction can make the work much easier.

EXAMPLE

Simplify: $\dfrac{1-\dfrac{1}{x+2}-\dfrac{2x-4}{x^2-4}}{x-\dfrac{x^3-4}{x^2-4}}$

The component fractions are $\dfrac{1}{x+2}$ and $\dfrac{1}{x^2-4}$ so the common denominator is x^2-4.

$$\dfrac{1-\dfrac{1}{x+2}-\dfrac{2x-4}{x^2-4}}{x-\dfrac{x^3-4}{x^2-4}}=\left(\dfrac{1-\dfrac{1}{x+2}-\dfrac{2x-4}{x^2-4}}{x-\dfrac{x^3-4}{x^2-4}}\right)\left(\dfrac{x^2-4}{x^2-4}\right)$$

$$=\dfrac{x^2-4-\dfrac{x^2-4}{x+2}-(2x-4)}{x(x^2-4)-(x^3-4)}$$

$$=\dfrac{x^2-4-(x-2)-2x+4}{x^3-4x-x^3+4}$$

$$=\dfrac{x^2-3x+2}{-4x+4}$$

Factor the numerator and denominator:

$$\dfrac{x^2-3x+2}{-4x+4}=\dfrac{(x-2)(x-1)}{-4(x-1)}=-\dfrac{x-2}{4}.$$

EXAMPLE

Simplify: $\dfrac{1+\dfrac{4x+4}{x^2+2x+4}}{1-\dfrac{x^3-x-12}{x^3-8}}$

Recall that $x^3-8=(x-2)(x^2+2x+4)$ so x^3-8 is the common denominator.

$$\left(\dfrac{1+\dfrac{4x+4}{x^2+2x+4}}{1-\dfrac{x^3-x-12}{x^3-8}}\right)\left(\dfrac{x^3-8}{x^3-8}\right)=\dfrac{x^3-8+(4x+4)(x-2)}{x^3-8-(x^3-x-12)}$$

$$=\dfrac{x^3-8+4x^2-4x-8}{-8+x+12}$$

$$=\dfrac{x^3+4x^2-4x-16}{x+4}$$

$$= \frac{\left(x^3 + 4x^2\right) - \left(4x + 16\right)}{x+4}$$

$$= \frac{x^2\left(x+4\right) - 4\left(x+4\right)}{x+4}$$

$$= \frac{\left(x^2 - 4\right)\left(x+4\right)}{x+4} = x^2 - 4$$

Determine the value of $1 + \cfrac{1}{1 + \cfrac{1}{1 + \cfrac{1}{1 + \cfrac{1}{1 + \cfrac{1}{2}}}}}$

This is a bit more complicated than the two-tiered complex fractions we've just examined. We will start at the base of the complex fraction and work our way up, $1 + \dfrac{1}{2} = \dfrac{3}{2}$. Rewrite the next tier, $1 + \cfrac{1}{1 + \cfrac{1}{2}}$, as $1 + \cfrac{1}{\cfrac{3}{2}} = 1 + \dfrac{2}{3} = \dfrac{5}{3}$.

Continue to make the revisions as you simplify each piece.

$$1 + \cfrac{1}{1 + \cfrac{1}{1 + \cfrac{1}{2}}} = 1 + \cfrac{1}{\cfrac{5}{3}} = 1 + \dfrac{3}{5} = \dfrac{8}{5}.$$

Next is $1 + \cfrac{1}{1 + \cfrac{1}{\cfrac{8}{5}}} = 1 + \cfrac{1}{1 + \cfrac{5}{8}} = 1 + \cfrac{1}{\cfrac{13}{8}} = 1 + \dfrac{8}{13} = \dfrac{21}{13}$

(Are you familiar with the Fibonacci sequence? If not, it begins with a pair of ones. From then on, the next term in the sequence is the sum of the previous two terms. $1 + 1 = 2$ so the first three terms of the sequence are 1, 1, 2. $1 + 2 = 3$ so the first four terms are 1, 1, 2, 3. Continue in the same manner to get:

1, 1, 2, 3, 5, 8, 13, 21, 34, 55, . . .

Did you notice that the partial sums of our work in evaluating the complex fraction are the ratio of consecutive numbers of the Fibonacci sequence?)

EXERCISE 6.4

Simplify each of the following complex fractions.

1. $1 + \cfrac{1}{1 - \cfrac{1}{1 + \cfrac{1}{1 - \cfrac{1}{1 + \cfrac{1}{1 - \cfrac{1}{2}}}}}}$

2. $\cfrac{x + 3 - \cfrac{7}{x - 3}}{x + 1 - \cfrac{5}{x - 3}}$

3. $\cfrac{5 + \cfrac{123x}{x^2 - 10}}{15 - \cfrac{56x - 170}{x^2 - 10}}$

4. $\cfrac{\cfrac{x + 1}{x + 6} - \cfrac{x - 1}{x - 6}}{\cfrac{x - 1}{x + 6} - \cfrac{x + 1}{x - 6}}$

5. $\cfrac{\cfrac{1}{5} + \cfrac{x + 11}{x^2 - 49}}{\cfrac{1}{10} - \cfrac{x - 1}{x^2 - 49}}$

Solving Rational Equations

When solving an equation of the form $\dfrac{x + 3}{6} + \dfrac{x - 1}{3} = 2$, one usually clears the fractions from the problem by multiplying both sides of the equation by the common denominator. (The solution to the equation is $x = \dfrac{11}{3}$.) The equations we'll look at in this section will also be rational with the exception that the denominators may also contain variable expressions. Our process will not change—we'll find the common denominator for all terms in the problems, multiply both sides of the equation by this value, and solve the resulting equation.

EXAMPLE

▶ Solve: $\dfrac{1}{x - 1} + \dfrac{1}{x + 2} = \dfrac{1}{2}$

▶ The common denominator is $2(x - 1)(x + 2)$. Multiply both sides of the equation by this value:

$$2(x - 1)(x + 2)\left(\frac{1}{x - 1} + \frac{1}{x + 2}\right) = \left(\frac{1}{2}\right)2(x - 1)(x + 2)$$

▶ Applying the distributive property, this becomes:

$$2(x - 1)(x + 2)\left(\frac{1}{x - 1}\right) + 2(x - 1)(x + 2)\left(\frac{1}{x + 2}\right) = (x - 1)(x + 2)$$

▶ Canceling terms, the equation now reads:

$$2(x+2)+2(x-1)=(x-1)(x+2)$$

▶ Apply the distributive property again to get:

$$2x+4+2x-2=x^2+x-2$$

▶ Gather all terms on one side of the equation:

$$x^2-3x-4=0$$

▶ Factor and solve:

$$(x-4)(x+1)=0 \text{ so } x=4,-1$$

Always check that the solutions found do, in fact, solve the problem.

EXAMPLE

▶ Solve:

$$\frac{x+5}{x+6}-\frac{x}{3(x-3)}=\frac{1}{4}$$

▶ The common denominator is $12(x+6)(x-3)$.

▶ Multiply both sides of the equation:

$$12(x+6)(x-3)\left(\frac{x+5}{x+6}-\frac{x}{3(x-3)}\right)=\left(\frac{1}{4}\right)12(x+6)(x-3)$$

▶ Apply the distributive property:

$$12(x+6)(x-3)\left(\frac{x+5}{x+6}\right)-12(x+6)(x-3)\left(\frac{x}{3(x-3)}\right)=3(x+6)(x-3)$$

▶ Cancel:

$$12(x-3)(x+5)-4x(x+6)=3(x+6)(x-3)$$

▶ Expand:

$$12x^2+24x-180-4x^2-24x=3x^2+9x-54$$

▶ Gather all terms to one side of the equation:

$$5x^2-9x-126=0$$

▶ Solve:

$$(5x+21)(x-6)=0 \text{ so } x=\frac{-21}{5},6$$

EXAMPLE

▶ Solve:

$$x - \frac{3}{x-4} = \frac{x-1}{4-x}$$

▶ Rewrite $\frac{x-1}{4-x}$ as $-\left(\frac{x-1}{x-4}\right)$. The common denominator is $x-4$.

▶ Multiply both sides of the equation by $x-4$:

$$(x-4)\left(x - \frac{3}{x-4}\right) = -\left(\frac{x-1}{x-4}\right)(x-4)$$

▶ Distribute:

$$x(x-4) - (x-4)\left(\frac{3}{x-4}\right) = -(x-1)$$

▶ Multiply:

$$x^2 - 4x - 3 = 1 - x$$

▶ Bring all terms to one side of the equation:

$$x^2 - 3x - 4 = 0$$

▶ Factor and solve:

$$(x-4)(x+1) = 0 \quad \text{so } x = 4, -1$$

▶ Did you check the solution? $x = 4$ is not a solution to this problem because it causes two of the denominators to be equal to zero. The only solution is $x = -1$. It is good practice to identify the domain of the problem as the first step in the process of solving equations.

Looking back at the problems just solved, the restrictions to the domain are $\frac{1}{x-1} + \frac{1}{x+2} = \frac{1}{2}, x \neq 1, -2$, for $\frac{x+5}{x+6} - \frac{x}{3(x-3)} = \frac{1}{4}, x \neq -6, 3$,

and $x - \frac{3}{x-4} = \frac{x-1}{4-x}, x \neq 4$.

EXAMPLE

▶ Solve:

$$\frac{2x}{x-2} + \frac{x-18}{(x-2)(x+2)} = 1$$

▶ The domain for this problem is $x \neq \pm 2$.

▶ The common denominator is $(x-2)(x+2)$.

► Multiply both sides of the equation by the common denominator:

$$(x-2)(x+2)\left(\frac{2x}{x-2}+\frac{x-18}{(x-2)(x+2)}\right)=1(x-2)(x+2)$$

► Distribute:

$$(x-2)(x+2)\left(\frac{2x}{x-2}\right)+(x-2)(x+2)\left(\frac{x-18}{(x-2)(x+2)}\right)=1(x-2)(x+2)$$

► Simplify:

$$2x(x+2)+x-18=x^2-4$$

► Distribute:

$$2x^2+4x+x-18=x^2-4$$

► Bring all terms to one side of the equation:

$$x^2+5x-14=0$$

► Factor and solve:

$$(x+7)(x-2)=0 \text{ so } x=-7,2$$

► Because 2 is not part of the domain, the correct solution to this equation is $x=-7$.

WORK PROBLEMS

How much time does it take to complete a job? The answer depends on how fast the person doing the job can work. The portion of the job performed is equal to the ratio of the time worked and the amount of time needed to complete the job. That is, if it takes you 4 hours to complete a task and you have worked on the task for 2 hours, half the task has been completed.

EXAMPLE

Colin can mow the lawn at his house in three hours. His brother Carson can mow the same lawn in four hours. Colin had been mowing for an hour when Carson told him that he had tickets for the game that afternoon. Colin explained that he had to finish the lawn before he could leave. Carson, knowing this, had borrowed a neighbor's mower, and together the two finished the lawn. How much time did it take for them to finish?

▶ Colin had already finished $\frac{1}{3}$ of the lawn when Carson arrived. If t represents the amount of time needed to complete the job, then Colin will mow an additional $\frac{t}{3}$ of the lawn while Carson will do $\frac{t}{4}$, and together they will complete the remaining $\frac{2}{3}$ of the lawn.

$$\frac{t}{3} + \frac{t}{4} = \frac{2}{3}$$

▶ Multiply both sides by the common denominator:

$$12\left(\frac{t}{3} + \frac{t}{4}\right) = \left(\frac{2}{3}\right)12$$

▶ This becomes $4t + 3t = 8$ or $7t = 8$ so they finish the lawn in $\frac{8}{7}$ hours.

EXAMPLE

▶ Colin and Carson's friends, Jamal and Allie, have a similar situation. They have plans to see a movie in the afternoon but need to get the lawn done first. It would take Allie 2 hours more than Jamal to mow the lawn if each did the mowing alone. However, working together, they can mow the lawn in 2.5 hours. How long would it take each to mow the lawn when working alone?

▶ If it takes Jamal h hours to mow the lawn, then it will take Allie $h + 2$ hours. When working together, the portion of the lawn Jamal mows is $\frac{2.5}{h}$ while Allie does $\frac{2.5}{h+2}$. Since they mow the entire lawn, together they complete 1 job. As a result, the equation is $\frac{2.5}{h} + \frac{2.5}{h+2} = 1$.

▶ Multiply by the common denominator to get:

$$2.5(h+2) + 2.5h = h(h+2)$$

▶ Distribute:

$$2.5h + 5 + 2.5h = h^2 + 2h$$

▶ Bring all terms to one side of the equation:

$$h^2 - 3h - 5 = 0$$

▶ Since this quadratic does not factor, use the quadratic formula to determine $h = \frac{3 + \sqrt{29}}{2}$ or 4.2 hours. (We reject $h = \frac{3 - \sqrt{29}}{2}$ as a solution because it is negative.) It would take Jamal 4.2 hours to mow the lawn alone, and it would take Allie 6.2 hours to mow the same lawn alone.

TRAVEL PROBLEMS

The basic equation for motion is $rt = d$ where r is the rate of travel, t is the time traveled, and d is the distance traveled. Problems involving wind and currents impact the rate of travel. Since we are still in the early process of studying mathematics, we make sure that the direction of the wind or current is parallel to the direction of the plane or boat.

EXAMPLE

A plane that travels 450 mph in still air makes a round trip in 9 hours. If the two towns from which the plane flew are 2,000 miles apart and the wind was constant through both legs of the trip, what was the speed of the wind?

The information in the problem relates the amount of time needed to make the round trip. Using the basic equation for motion, we can write that time $= \dfrac{\text{distance}}{\text{rate}}$. The equation is $\dfrac{2{,}000}{450+w} + \dfrac{2{,}000}{450-w} = 9$ where $450 + w$ represents the speed of the plane with the wind at its back (thus increasing the overall speed) and $450 - w$ represents the speed of the plane with the wind in its face (thus decreasing the overall speed) and each fraction representing the time for that leg of the trip. Multiply by the common denominator:

$$2{,}000(450 - w) + 2{,}000(450 + w) = 9(450 + w)(450 - w)$$

Simplify:

$$900{,}000 - 2{,}000w + 900{,}000 + 2{,}000w = 9(202{,}500 - w^2)$$

which becomes:

$$1{,}800{,}000 = 9(202{,}500 - w^2).$$

Divide by 9:

$200{,}000 = 202{,}500 - w^2$ which becomes $w^2 = 2{,}500$ so that $w = 50$ (ignore the -50 in this application). The wind was blowing at 50 mph.

Minimum and Maximum Values

The availability of graphing technology makes the solution of extreme value problems easier to do. The emphasis is on writing the correct functions to solve the problem.

A manufacturer is making a rectangular container to hold 500 cubic centimeters (cc). The base of the container is to be a square, while the sides are to be rectangles. Because of the stress on the base of the box, it is to be made of a sturdier material than the vertical sides and top of the box. The cost of the material for the base of the box is $0.005 per square centimeter, while the cost of the materials for the rest of the box is $0.003 per square centimeter. What are the dimensions of the box with minimum cost?

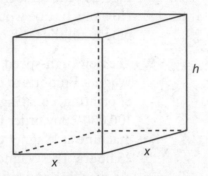

The area of each side panel is xh sq. cm., and the area of the top and bottom of the box is x^2 sq. cm. The volume of the box is given by the equation $x^2h = 500$, and the cost, C, is based on the area of the four side panels and the area of the top and bottom of the box. The cost of making the box is given by the equation $C = 0.005x^2 + 0.003(4xh + x^2)$. Simplifying this equation, we get $C = 0.008x^2 + 0.012xh$. Using $x^2h = 500$, we get $h = \dfrac{500}{x^2}$. Substituting this into the cost equation, we have

$$C = 0.008x^2 + 0.012x\left(\frac{500}{x^2}\right) = 0.008x^2 + \frac{6}{x}.$$ This is a function whose graph is most likely not familiar to us. Use the graphing calculator to sketch the graph of the cost function and then use the Minimum option under your Graph menu to determine that the base of the box measures 7.21 cm by 7.21 cm. Substitute 7.21 for x in the equation $x^2h = 500$ and the height of the box is 9.62 cm.

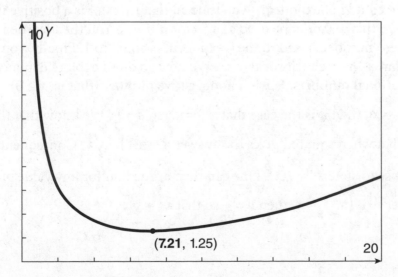

EXERCISE 6.5

Solve each equation.

1. $\dfrac{4}{4-x} - \dfrac{x^2}{4-x} = -5$

2. $\dfrac{x-5}{x-2} + \dfrac{2}{x-7} = \dfrac{31}{(x-2)(x-7)}$

3. $\dfrac{4x-2}{x^2-1} - \dfrac{1}{x} = \dfrac{66}{x(x^2-1)}$

4. $\dfrac{x+1}{2x} + \dfrac{x-4}{x+2} = \dfrac{5}{x}$

5. $\dfrac{4}{x^2-8x+12} - \dfrac{x}{x-2} = \dfrac{1}{x-6}$

6. $\dfrac{x+1}{3x-4} + \dfrac{2}{x-3} = \dfrac{6x+4}{(3x-4)(x-3)}$

7. $\dfrac{x+3}{x^2-1} - \dfrac{2x-1}{7x+2} = \dfrac{3x-5}{7x+2}$

8. An inlet water pipe can fill a tank in 6 hours. An hour after this pipe has been open a second inlet pipe is opened and, together, the tank is filled in 3 more hours. How much time would it take for the second pipe to fill the tank by itself?

9. The new high-speed printing machine at Wagner Emporium can complete the job of printing, collating, and stapling the 100,000-copy order in 15 hours. The older machine at Wagner can do the same job in 25 hours. The Wagners start the job with the old machine and, after 3 hours, put the new machine on the job as well. How long will it take to finish the job?

10. Max and Sally are taking a canoe ride on a stream that has a steady current of 2 mph. They complete a 10-mile trip downstream and the return trip upstream in 6 hours 40 minutes. How fast can Max and Sally paddle a canoe in water with no current?

Rational Exponents

In your first study of exponents, you learned that that if n is a positive integer, x^n indicated that x should be used as a factor n times. You then learned that any nonzero number raised to the 0 exponent is equal to 1. Finally, you learned that x^{-n} is the reciprocal of x^n. We now go on to explore exponents that are rational numbers. If x is a nonnegative number (that is, $x \geq 0$) and $(x^a)^2 = x$, then it is the case that $x^{2a} = x^1$ so $a = \dfrac{1}{2}$. We know that the number that when squared gives the answer x must be $\sqrt{x}$. Consequently, it makes sense to state $x^{\frac{1}{2}} = \sqrt{x}$. One can then argue that for any value of x and any integer n, if $(x^a)^n = x$ then $a = \dfrac{1}{n}$ so that $x^{\frac{1}{n}} = \sqrt[n]{x}$.

EXAMPLE

Simplify $64^{\frac{1}{3}}$.

By rule:

$$64^{\frac{1}{3}} = \sqrt[3]{64} = 4.$$

If $x^{\frac{1}{n}} = \sqrt[n]{x}$ then it makes a great deal of sense (in order to be consistent with the rules of exponents) that $x^{\frac{m}{n}} = \left(x^{\frac{1}{n}} \right)^{m} = \left(\sqrt[n]{x} \right)^{m}$.

EXAMPLE

Simplify $64^{\frac{2}{3}}$ and $64^{\frac{-2}{3}}$.

$$64^{\frac{2}{3}} = \left(64^{\frac{1}{3}} \right)^{2} = 4^{2} = 16 \text{ and } 64^{\frac{-2}{3}} = \left(64^{\frac{2}{3}} \right)^{-1} = 16^{-1} = \frac{1}{16}.$$

EXAMPLE

Simplify $\left(27x^{6}y^{12} \right)^{\frac{4}{3}}$.

Apply the rules for exponents and rewrite:

$$\left(27x^{6}y^{12} \right)^{\frac{4}{3}} \text{ as } (27)^{\frac{4}{3}} \left(x^{6} \right)^{\frac{4}{3}} \left(y^{12} \right)^{\frac{4}{3}} = 81x^{8}y^{16}$$

EXAMPLE

Simplify $\left(\dfrac{36x^{4}y^{-5}}{4x^{-2}y^{-3}} \right)^{\frac{-3}{2}}$ and write your answer without negative exponents.

Simplify the terms inside the parentheses:

$$\left(\frac{36x^{4}y^{-5}}{4x^{-2}y^{-3}} \right)^{\frac{-3}{2}} = \left(\frac{9x^{6}}{y^{2}} \right)^{\frac{-3}{2}}$$

Rewrite with the positive exponent $\dfrac{3}{2}$:

$$\left(\frac{9x^{6}}{y^{2}} \right)^{\frac{-3}{2}} = \left(\frac{y^{2}}{9x^{6}} \right)^{\frac{3}{2}}$$

Apply the rules of exponents:

$$\left(\frac{y^{2}}{9x^{6}} \right)^{\frac{3}{2}} = \frac{\left(y^{2} \right)^{\frac{3}{2}}}{9^{\frac{3}{2}} \left(x^{6} \right)^{\frac{3}{2}}} = \frac{y^{3}}{27x^{9}}.$$

EXERCISE 6.6

Simplify each of the following. Assume all literal values are positive. Write answers without negative exponents.

1. $\left(128x^{14}y^{21}\right)^{\frac{2}{7}}$

2. $\left(\dfrac{81a^{12}b^7}{16a^{-4}b^{11}}\right)^{\frac{3}{4}}$

3. $\dfrac{\left(125x^9z^8\right)^{\frac{2}{3}}}{\left(64x^{12}z^2\right)^{\frac{1}{6}}}$

4. $\left(100b^4c^6\right)^{\frac{1}{2}}\left(512b^{-6}c^9\right)^{\frac{1}{3}}$

5. $\left(81m^{12}n^{24}\right)^{\frac{1}{2}}+\left(9m^4n^8\right)^{\frac{3}{2}}$

Simplifying Irrational Expressions

Rational exponents can be used to simplify more complicated irrational expressions. You can help yourself by taking the time to learn at least the first 5 powers of the numbers 2 through 5 and the first 4 powers of 6 through 9.

n	n^2	n^3	n^4	n^5
2	4	8	16	32
3	9	27	81	243
4	16	64	256	1,024
5	25	125	625	3,125
6	36	216	1,296	—
7	49	243	2,401	—
8	64	512	4,096	—
9	81	729	6,561	—

EXAMPLE

Simplify: $\sqrt[4]{48}$.

You know that $48 = 16 \times 3$ so $\sqrt[4]{48} = \sqrt[4]{16}\,\sqrt[4]{3} = 2\sqrt[4]{3}$.

EXAMPLE

Simplify: $\sqrt[3]{486}$.

$486 = 243 \times 2$ so $\sqrt[3]{486} = \sqrt[3]{243}\,\sqrt[3]{2} = 7\sqrt[3]{2}$

EXAMPLE

▶ Simplify $\sqrt[5]{64x^7y^{10}}$.

▶ Rewrite the radical as an expression with rational exponents:

$$\sqrt[5]{64x^7y^{10}} = \left(64x^7y^{10}\right)^{\frac{1}{5}}$$

▶ This is equal to:

$$\left(2^6\right)^{\frac{1}{5}}\left(x^7\right)^{\frac{1}{5}}\left(y^{10}\right)^{\frac{1}{5}} = 2^{\frac{6}{5}}x^{\frac{7}{5}}y^2 = 2(2)^{\frac{1}{5}}x\left(x^2\right)^{\frac{1}{5}}y^2$$

▶ These last two expressions are acceptable as answers. However, if the directions are to write the answer as a radical expression (or the choices for a multiple choice test are written as radicals):

$$2(2)^{\frac{1}{5}}x\left(x^2\right)^{\frac{1}{5}}y^2 = 2xy^2\sqrt[5]{2x^2}$$

▶ Simplify $\sqrt[5]{64x^8y^{13}}$.

▶ Working with fifth powers:

$$\sqrt[5]{64x^8y^{13}} = \sqrt[5]{32x^5y^{10}}\,\sqrt[5]{2x^3y^3} = 5xy^2\sqrt[5]{2x^3y^3}$$

▶ Simplify $\dfrac{\sqrt[3]{192x^5y^7}}{\sqrt[4]{48x^6y^{11}}}$.

▶ Use rational exponents to simplify each expression:

$$\frac{\sqrt[3]{192x^5y^7}}{\sqrt[4]{48x^6y^{11}}} = \frac{\left(192x^5y^7\right)^{\frac{1}{3}}}{\left(48x^6y^{11}\right)^{\frac{1}{4}}}$$

$$= \frac{\left(64x^3y^6\right)^{\frac{1}{3}}\left(3x^2y\right)^{\frac{1}{3}}}{\left(16x^4y^8\right)^{\frac{1}{4}}\left(3x^2y^3\right)^{\frac{1}{4}}}$$

$$= \frac{4xy^2\left(3x^2y\right)^{\frac{1}{3}}}{2xy^2\left(3x^2y^3\right)^{\frac{1}{4}}} = \frac{2\left(3x^2y\right)^{\frac{1}{3}}}{\left(3x^2y^3\right)^{\frac{1}{4}}}$$

$$= \frac{2\left(3^{\frac{1}{3}}x^{\frac{2}{3}}y^{\frac{1}{3}}\right)}{\left(3^{\frac{1}{4}}x^{\frac{1}{2}}y^{\frac{3}{4}}\right)}$$

We know that $\dfrac{3^{\frac{1}{3}}}{3^{\frac{1}{4}}}=3^{\frac{1}{3}-\frac{1}{4}}=3^{\frac{1}{12}}$ because the rules of exponents tell us that when the bases are the same, one subtracts exponents when doing a division problem. Therefore:

$$\frac{3^{\frac{1}{3}}x^{\frac{2}{3}}y^{\frac{1}{3}}}{3^{\frac{1}{4}}x^{\frac{1}{2}}y^{\frac{3}{4}}}=\frac{3^{\frac{1}{12}}x^{\frac{1}{6}}}{y^{\frac{5}{12}}}$$

Finally:

$$\frac{\sqrt[3]{192x^5y^7}}{\sqrt[4]{48x^6y^{11}}}=\frac{2\left(3^{\frac{1}{12}}x^{\frac{1}{6}}\right)}{y^{\frac{5}{12}}}$$

EXERCISE 6.7

Simplify each of the following. Assume all literal values are positive.

1. $\sqrt[3]{1,000x^9y^{11}}$

2. $\sqrt[4]{32x^9z^{21}}$

3. $\left(\sqrt[3]{3x^2y^4}\right)\left(\sqrt[3]{72x^4y^{-7}}\right)$

4. $\dfrac{\sqrt{32a^9b^7}}{\sqrt[4]{32a^6b^{14}}}$

Solving Irrational Equations

We solve irrational equations, whether written with the radical or with rational exponents, by isolating the irrational expression and then raising both sides of the equation by the appropriate power to remove the radical (or the exponent).

EXAMPLE

Solve: $\sqrt{2x+3}-5=8$

Isolate the radical by adding 5 to both sides of the equation:

$$\sqrt{2x+3}=13$$

Square both sides of the equation:

$$\left(\sqrt{2x+3}\right)^2=13^2 \text{ becomes } 2x+3=169.$$

Solve the linear equation:

$$2x+3=169\Rightarrow2x=166\Rightarrow x=83$$

EXAMPLE

Solve $\sqrt[3]{x^2 - 33} + 5 = 3$

Isolate the radical:

$$\sqrt[3]{x^2 - 33} = -2$$

Cube both sides of the equation:

$$\left(\sqrt[3]{x^2 - 33}\right)^3 = (-2)^3 \text{ becomes } x^2 - 33 = -8.$$

Add 33 to both sides of the equation and solve:

$$x^2 - 33 = -8 \Rightarrow x^2 = 25 \Rightarrow x = \pm 5$$

EXAMPLE

Solve: $\sqrt{2x + 11} + x = 12$

Yes, this is different with there being a variable in the radicand (the expression inside the radical) and outside. However, the process does not change.

Isolate the radical:

$$\sqrt{2x + 11} + x = 12 \text{ becomes } \sqrt{2x + 11} = 12 - x.$$

Remove the radical by squaring both sides of the equation:

$$\left(\sqrt{2x + 11}\right)^2 = (12 - x)^2 \text{ becomes } 2x + 11 = 144 - 24x + x^2$$

Set one side of the equation equal to zero:

$$x^2 - 26x + 133 = 0$$

Factor (or use the quadratic formula) to solve for x:

$$(x - 7)(x - 19) = 0 \text{ and } x = 7, 19$$

Check your answers: $x = 7$:

$$\sqrt{2(7) + 11} + 7 = 12 \text{ becomes } \sqrt{25} + 7 = 12 \text{ so that } 5 + 7 = 12. \text{ Correct.}$$

Check $x = 19$:

$$\sqrt{2(19) + 11} + 19 = 12 \text{ becomes } \sqrt{49} + 19 = 12 \text{ so that } 7 + 19 = 12.$$
Not correct.

Therefore, $x = 7$ is the solution (and $x = 19$ is called an **extraneous root**).

EXAMPLE

Solve: $\sqrt{2x-3}+\sqrt{3x+7}=12$

The first step is to isolate the radical. But there are two radicals in the problem! What to do? Easy, you decide.

Isolate $\sqrt{2x-3}$

$$\sqrt{2x-3}=12-\sqrt{3x+7}$$

Isolate $\sqrt{3x+7}$

$$\sqrt{3x+7}=12-\sqrt{2x-3}$$

Square both sides of the equation:

$$\left(\sqrt{2x-3}\right)^2=\left(12-\sqrt{3x+7}\right)^2$$

Square both sides of the equation:

$$\left(\sqrt{3x+7}\right)^2=\left(12-\sqrt{2x-3}\right)^2$$

Expand:

$$2x-3=144-24\sqrt{3x+7}+3x+7$$

Expand:

$$3x+7=144-24\sqrt{2x-3}+2x-3$$

Simplify:

$$2x-3=3x+151-24\sqrt{3x+7}$$

Simplify:

$$3x+7=2x+141-24\sqrt{2x-3}$$

Isolate the radical (again):

$$24\sqrt{3x+7}=x+154$$

Isolate the radical (again):

$$24\sqrt{2x-3}=-x+134$$

Square both sides of the equation:

$$\left(24\sqrt{3x+7}\right)^2=\left(x+154\right)^2$$

$$576(3x+7)=x^2+308x+23{,}716$$

$$1{,}728x+4{,}032=x^2+308x+23{,}716$$

Square both sides of the equation:

$$\left(24\sqrt{2x-3}\right)^2=\left(-x+134\right)^2$$

$$576(2x-3)=x^2-268x+17{,}956$$

$$1{,}152x-1{,}728=x^2-268x+17{,}956$$

Bring all terms to one side of the equation:

$$x^2-1{,}420x+19{,}684=0$$

Bring all terms to one side of the equation:

$$x^2-1{,}420x+19{,}684=0$$

Too big to factor, use the quadratic formula:

$$x=\frac{1{,}420\pm\sqrt{1{,}937{,}664}}{2}$$

$$x=14,1406$$

This is the same equation from before.

Therefore, $x = 14$. (You can show that 1,406 is an extraneous root.)

It isn't often that you will see problems with indices (the number which indicates the root in question) because the algebra gets particularly ugly. Having said that:

EXAMPLE

Solve: $\sqrt[3]{x^2-17}=x-5$

We can cube both sides of the equation to get:

$$x^2-17=x^3-15x^2+75x-125$$

which becomes:

$$x^3-16x^2+75x-108=0$$

Rather than try to factor this difficult problem, use your graphing utility to graph each of the original functions and find the point of intersection.

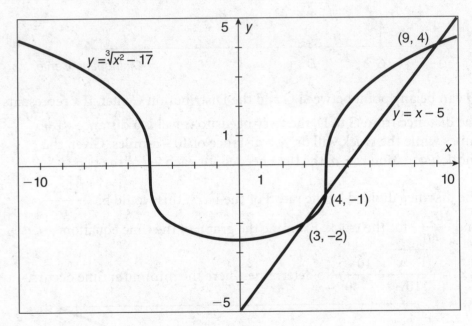

Use the Intersect command to show that the solution to the problem is $x=3, 4, 9$.

EXAMPLE

A factory and distribution center lie on opposite sides of a river that is 1 mile wide (as shown in the diagram).

Factory

River

Distribution Center

The company needs to move its product from the factory to the distribution center and wishes to do so in the least amount of time. A ferry can carry a truck from the factory side of the river to the opposite shore at a maximum speed of 10 mph. The truck can disembark and travel on a road to the distribution center at a speed of 40 mph. If the distribution center is 10 miles downstream from the factory and the time it takes for the truck to embark and disembark is negligible, where should the company build a dock on the distribution center side of the river to minimize the time needed to transport the product?

Let D represent the point where the dock should be built and G represent the point directly across the river from the factory.

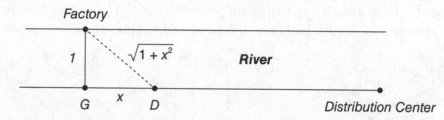

D can be any point between G and the Distribution Center. If x represents the distance from G to D, the ferry needs to travel is a distance $\sqrt{1+x^2}$ miles while the truck will drive a distance of $10 - x$ miles. Given the motion equation $rt = d$, the time needed for each leg of the trip is the distance divided by the rate. For the ferry, this would be $\dfrac{\sqrt{1+x^2}}{10}$ and $\dfrac{10-x}{40}$ for the truck. Examine the graph of the time equation

$$t(x) = \frac{\sqrt{1+x^2}}{10} + \frac{10-x}{40}$$ to determine where the minimum time occurs.

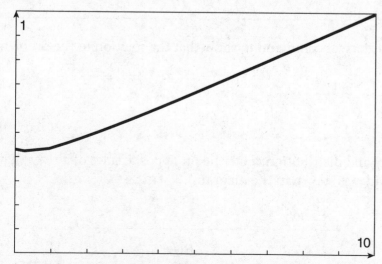

Use the minimum feature from the Graph menu on your calculator to determine that the dock should be built 0.258 miles from point G.

EXERCISE 6.8

Solve each of the following equations.

1. $\sqrt[4]{5x+1}=2$

2. $\sqrt{21-5x}=6$

3. $\sqrt{7x-3}=5$

4. $\sqrt{3x+6}+4=x$

5. $\sqrt{11-2x}-x=12$

6. $\sqrt{x^2+9}+3=2x$

7. $\sqrt{8x+1}+\sqrt{39-x}=11$

8. $\sqrt{6x+27}-\sqrt{x+7}=5$

9. $\sqrt[3]{4x+3}=3$

10. $\sqrt[3]{8x-13}-\sqrt[3]{x+3}=1$

Solving Rational Inequalities

We know to solve the equation $\dfrac{3x-8}{2x+7}=0$ that we simply multiply both sides of the equation by the denominator and solve the resulting linear equation.

However, we do not have that luxury when solving the inequality $\dfrac{3x-8}{2x+7}>0$.

Why? For some values of x the denominator is positive so there is no issue, but for other values of x the denominator is negative and we would need to reverse the inequality because of the impact on multiplying both sides of an inequality by a negative. So, what to do?

We'll start as we have a number of times in this book and look at the graph of the problem. From there, we'll develop a graph-less process to get to the

solution. Let's take a look at the graph of $y=\dfrac{3x-8}{2x+7}$.

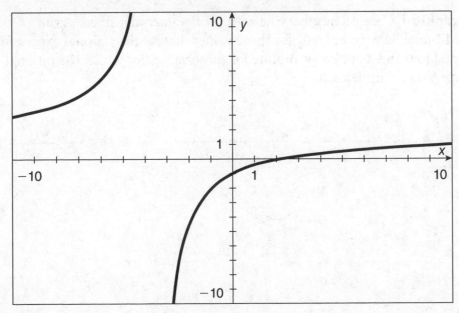

We can see that the graph crosses the x-axis at $x = \frac{8}{3}$ and that it fails to exist at $x = \frac{-7}{2}$ (both points can be deduced from looking at the equation). We are going to take advantage of a very simple concept: given any real number y, it is either negative, zero, or positive.

As we look at the graph, we see that there is only one place where the value of $y = 0$, and that is when $x = \frac{8}{3}$. All other values of y are either positive or negative. For all values of $x > \frac{8}{3}$, $y > 0$. It is also true that $y > 0$ whenever $x < \frac{-7}{2}$. For all values of x for which $\frac{-7}{2} < x < \frac{8}{3}$, y is negative. We now know the solution to $\frac{3x-8}{2x+7} > 0$ is $x < \frac{-7}{2} \cup x > \frac{8}{3}$.

To shift this discussion from a strictly graphical approach to an algebraic/numerical/graphical approach, we need to realize the points that are critical to us are those points from which the rational expression is equal to 0 and those points for which the expression fails to exist. That is, we need to determine when the numerator equals zero and when the denominator equals zero. We know that for every other value of x, the expression will either be positive or negative. We will use a **signs analysis** to examine the value of the dependent variable.

<table>
<tr>
<td>

Signs analysis is concerned only with the sign of an outcome, not the magnitude. When using signs analysis, avoid getting wrapped up in difficult computations and concentrate on simple things like a positive times a positive divided by negative is a negative result.

</td>
<td>

EXAMPLE

Solve: $\dfrac{x^2 - 3x - 4}{x - 2} > 0$

Algebraic. Factor the numerator: $\dfrac{(x-4)(x+1)}{x-2} > 0$

Set each factor equal to 0: $x = 4$, -1 sets the numerator (and expression) equal to 0; $x = 2$ sets the denominator equal to zero (and is the value for which the expression is undefined).

Graphical. Create a number line with these values identified. As an additional help, write "= 0" for those values that set the rational expression equal to 0 and a dotted asymptote for those values for which the rational expression is undefined.

</td>
</tr>
</table>

▶ **Numerical.** Pick any number in the leftmost set of numbers (in this case, to the left of –1) and examine the sign of each factor in the rational expression. For example, using $x = -2$, $\dfrac{(x-4)(x+1)}{x-2}$ becomes $\dfrac{(-2-4)(-2+1)}{-2-2}$ and the signs are $\dfrac{(-)(-)}{-}$ and that leads to a negative result. The expression $\dfrac{(x-4)(x+1)}{x-2}$ will be a negative value when evaluated with any value of x smaller than –1. Choose $x = 0$ for the value of x between –1 and 2, choose $x = 3$ for the interval from $x = 2$ to $x = 4$, and choose $x = 5$ for a value of $x > 4$. (Remember, you can choose *any* value in the interval. The values chosen seemed easiest to the author.) Note the results on the number line with the signs of each factor above the number line and the sign of the expression below the number line.

$$\underset{-}{\dfrac{(-)(-)}{-}} \quad = 0 \qquad \underset{-}{\dfrac{(-)(+)}{-}} \; \vdots \; \underset{+}{\dfrac{(-)(+)}{+}} \quad = 0 \qquad \underset{+}{\dfrac{(+)(+)}{+}}$$

| $-$ | -1 | $+$ | 2 | $-$ | 4 | $+$ |

▶ The value of the expression $\dfrac{(x-4)(x+1)}{x-2}$ is positive when $-1 < x < 2$ or $x > 4$.

EXAMPLE

▶ Solve: $\dfrac{2x^2 - 5x - 3}{x^2 - 4} \le 0$

▶ Factor the terms in the rational expression:

$$\dfrac{(2x+1)(x-3)}{(x-2)(x+2)} \le 0$$

▶ Determine the values for which the rational expression equals 0 $\left(x = \dfrac{-1}{2}, 3\right)$ and those for which the rational expression is undefined $(x = \pm 2)$.

▶ Create the number with these values and appropriate designations assigned.

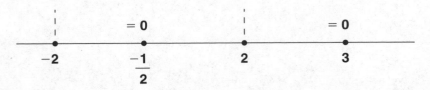

The result of the signs analysis for each interval, with the value chosen to test the interval, is shown on the number line.

$$\frac{(-)(-)}{(-)(-)} \quad \vdots \quad \frac{(+)(-)}{(-)(-)} \quad \cdot 0 \quad \frac{(+)(-)}{(-)(+)} \quad \vdots \quad \frac{(+)(-)}{(+)(+)} \quad \cdot 0 \quad \frac{(+)(+)}{(+)(+)}$$

x = −3 −2 x = −1 $\frac{-1}{2}$ x = 0 2 x = 2.5 3 x = 4
+   −   +   −   +

The solution to $\dfrac{2x^2 - 5x - 3}{x^2 - 4} \le 0$ is $-2 < x \le \dfrac{-1}{2} \cup 2 < x \le 3$.

Warning: Do not assume from these two examples that once you've determined the sign for the leftmost interval the signs for the remaining intervals will alternate.

Solve: $\dfrac{x^2 - x - 6}{(x-1)^2} \ge 0$

Factor:

$$\frac{(x-3)(x+2)}{(x-1)^2}$$

The numerator equals zero when $x = -2, 3$ while the denominator equals zero when $x = 1$. The number line and signs analysis are

$$\frac{(-)(-)}{(+)} \quad = 0 \quad \frac{(-)(+)}{(+)} \quad \vdots \quad \frac{(-)(+)}{(+)} \quad = 0 \quad \frac{(+)(+)}{(+)}$$

x = −3 −2 x = 0 1 x = 2 3 x = 4
+   −   −   +

The solution to $\dfrac{x^2 - x - 6}{(x-1)^2} \ge 0$ is $x \le -2$ or $x \ge 3$.

EXERCISE 6.9

Solve each inequality.

1. $\dfrac{x^2 - 4}{x - 1} < 0$

2. $\dfrac{x^2 - 4x - 5}{x^2 - 4} > 0$

3. $\dfrac{x^3 - 4x^2 - 12x}{x^2 - 2x - 8} \leq 0$

4. $\dfrac{x^4 - 16x^2}{x^2 - 9} \geq 0$

5. $\dfrac{3x^2 - 4x - 15}{x^2 - x} > 0$

Variation: Direct, Inverse, and Joint

You have been working with proportions for a number of years now. Solving an equation such as $\dfrac{260}{4} = \dfrac{x}{7}$ is an easy problem for you to do. Here's a problem for which the proportion might be used. If Carson can drive 260 miles in 4 hours, how far can he drive (assuming the same speed) in 7 hours? The item that is important here is the units of each fraction. The term $\dfrac{260 \text{ miles}}{4 \text{ hours}}$ gives the average speed of the problem. Consequently, the term $\dfrac{x}{7}$ must be in the same units. In a proportion, each fraction represents a constant. That is, $\dfrac{a}{b} = k$ (or, $a = kb$), where k is a constant. Each equation represents a direct variation. There are many examples of direct variations in your life: mph, mpg, cost per kilogram, dollars per hour.

Hooke's Law. When a weight is attached to the end of a spring, the length the spring stretches varies directly with the weight of the spring.

EXAMPLE

▶ If a weight of 20 kg stretches a spring 45 cm, what weight is needed to stretch a spring 60 cm?

▶ According to the statement of the problem, $l = kw$. Using the initial conditions, $45 = 20k$ so $k = \dfrac{9}{4}$ (cm/kg). Solve the equation $60 = \dfrac{9}{4} w$ to determine that $w = \dfrac{240}{9} = \dfrac{80}{3}$ kg.

Note: Not all direct variations involve linear expressions.

Free Fall. When an object falls from a height, the distanced fallen varies directly with the square of the time the object has been falling. A parachutist in free fall (the parachute is still closed) falls 64 feet in 2 seconds.

▶ How much time is needed for her to fall 250 feet?

▶ According to the statement of the problem:

$$d = kt^2$$

▶ Using the initial conditions:

$$64 = k(2)^2 \text{ so that } k = 16 \text{ ft/sec}^2$$

▶ Solving the problem:

$$250 = 16t^2 \text{ yields } t = 3.953 \text{ seconds (rounded to 3 decimal places).}$$

Kepler's Third Law. Kepler's third law of planetary motion states that the square of the time required for a planet to make one revolution about the sun varies directly as the cube of the average distance of the planet from the sun.

▶ If you assume that Venus is 0.72 times as far from the sun as is Earth, find the approximate length of a Venusian year. (The scale for distance is the astronomical unit (AU), the average distance Earth is from the sun.)

▶ According to the **statement of the problem**:

$$t^2 = kd^3$$

▶ Using Earth as the original condition, both t and d equal 1, so $k = 1$. Solving the problem for Venus:

$$t^2 = (1)(0.72)^3 \text{ and } t = (0.72)^{\frac{3}{2}} = 0.61 \text{ years (or 223 days).}$$

In **joint variation** one variable varies directly with the product of two or more variables.

EXAMPLE

G varies jointly with m and n. If $G = 300$ when $m = 20$ and $n = 3$, find the value of G when $m = 60$ and $n = 5$.

According to the statement of the problem, $G = kmn$. Using the initial conditions:

$$300 = k(20)(3) \text{ so that } k = 5$$

Solving the problem:

$$G = 5(60)(5) = 1500$$

EXAMPLE

Given V varies jointly with the square of x and the cube of y. If $V = 5{,}000$ when $x = 4$ and $y = 5$, find the value of V when $x = 6$ and $y = 8$.

According to the statement of the problem:

$$V = kx^2 y^3$$

Using the initial conditions:

$$5{,}000 = k(4)^2(5)^3$$

This becomes:

$$5{,}000 = k(16)(125) \text{ so that } k = 2.5$$

Solving the problem:

$$V = 2.5(6)^2(8)^3 = 46{,}080$$

Inverse variation differs from direct variation in that rather than having the ratio of variable be constant, the product of the variables is constant.

EXAMPLE

Given that x and y vary inversely and that when $x = 12$, $y = 10$. Determine the value of y when $x = 15$.

According to the statement of the problem:

$$xy = k$$

Using the initial conditions:

$$(12)(10) = k \text{ so } k = 120$$

▸ Solving the problem:

$$15y = 120 \text{ so } y = 8.$$

Fixed Area. In a rectangle with fixed area, the *length* and *width* vary inversely.

▸ If the width is 25 cm when the length is 40 cm, determine the width of the rectangle when the length is 60 cm.

▸ According to the statement of the problem, (length)(width) = k. Using the initial conditions, (25)(40) = k so that k = 1,000 sq. cm. Solving the problem, (60)(*width*) = 1,000 so *width* = $\dfrac{50}{3}$ cm.

Teeter-Totter. In order to balance two weights on a teeter-totter (or any type of fulcrum and lever), the *distance* from the fulcrum and the *weight* vary inversely.

▸ Colin weighs 40 pounds and sits 45 inches from the fulcrum. How far from the fulcrum should his younger brother Carson sit given that Carson weighs 30 pounds?

▸ According to the statement of the problem, (*length*)(*weight*) = k. Solving the problem, (40)(45) = k = (30)(*length*) or (40)(45) = (30)(*length*) so *length* = 60 inches.

There are a large number of problems that involve a combination of these variations.

Ideal Gas Law. The volume (V) of gas varies jointly with the number of moles of gas (n) and the temperature (T) of the gas (measured in degrees Kelvin) and inversely with the Pressure (P). One mole of gas has a volume of 40 cubic centimeters (cc) when the temperature is 293°K and the pressure is 2 Pascals.

What will the volume of one mole of gas be if the temperature is 300°K and the pressure is 3 Pascals?

According to the statement of the problem, $PV = nkT$. Using the initial conditions, $(2)(40) = (1)k(293)$ gives $k = \dfrac{80}{293}$. Solving the problem, $3V = (1)\left(\dfrac{80}{293}\right)(300)$ so that $V = 27.3$ cc.

Electrical Resistance. The number of ohms in the electrical resistance, Ω, in a wire varies directly with length, l, and inversely with the square of the diameter, d.

If the resistance in a wire with diameter 1 inch and length 10 feet is 5 ohms, what is the number of ohms in a wire 25 feet long with a diameter of 2 inches?

According to the statement of the problem, $\Omega d^2 = kl$. Using the initial conditions, $5(1)^2 = k(10)$ so that $k = \dfrac{1}{2}$. Solving the problem, $\Omega(2)^2 = \dfrac{1}{2}(25) = 3.125$ ohms.

Centrifugal Force. The centrifugal force of an object moving in a circle varies jointly with the radius of the circular path and the mass of the object and inversely as the square of the time it takes to move about one full circle.

If the force created by an object weighing 5 grams traveling in a circular path with radius 20 meters with a period of 2.5 seconds is 1,500 dynes, what force is created when an object with mass 8 g travels in a circle with radius 25 meters with a period of 2 seconds?

The period is the time it takes to complete a revolution around the circle. According to the statement of the problem, $p^2F = krm$. Using the initial conditions of the problem, $(2.5)^2(1,500) = k(20)(5)$ so that $k = 93.75$ Solving the problem, $(2)^2F = 93.75(25)(8)$ becomes $F = 4,678.5$ dynes.

EXERCISE 6.10

1. Use Kepler's Third Law of Planetary Motion to determine the length of the Jovian years, assuming that Jupiter is 5.2 times as far from the sun as is the earth.

2. Apply Hooke's Law to a spring in which a weight of 50 kg stretches a spring 125 cm. What weight is needed to stretch a spring 300 cm?

3. On a teeter-totter, Will weighs 162 pounds and sits 10 inches from the fulcrum. How far from the fulcrum should his daughter Ainsley sit from the fulcrum given that Ainsley weighs 30 pounds?

4. If the centrifugal force created by an object weighing 25 grams traveling in a circular path with radius 40 meters with a period of 2 seconds is 5,000 dynes, what force is created when an object with mass 30 g travels in a circle with radius 75 meters with a period of 1.5 seconds?

5. If two moles of gas has a volume of 400 cubic centimeters (cc) when the temperature is 300°K and the pressure is 1.5 Pascals, what will the volume of three moles of gas be when the temperature is 450°K and the pressure is 2 Pascals?

CHAPTER 7

Exponential and Logarithmic Functions

Population growth, radioactive decay, and compound interest are a few examples of exponential functions. Unlike linear growth in which the change from one period to the next is accomplished by the addition (or subtraction) of some constant, the change in value from one period to the next in exponential growth is by a constant factor. Albert Einstein is credited with saying that the most impactful mathematical creation was compound interest. In this chapter we will examine exponential functions, their inverses—logarithmic functions—and applications of both.

Exponential Functions

The basic form of the exponential function is $f(x) = (b)^x$ where b is a nonnegative constant and $b \neq 1$ (with 1 raised to any power being equal to 1, having $b = 1$ would just create a horizontal line). If $b > 1$, the graph would increase from left to right, while if $0 < b < 1$, the graph would decrease from left to right. The domain of the function is the set of real numbers, and the range of the basic exponential function is $y > 0$. The y-intercept for this basic is 1.

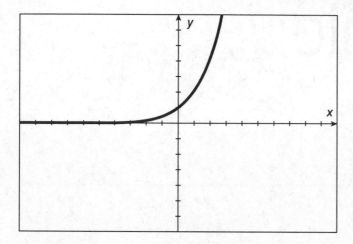
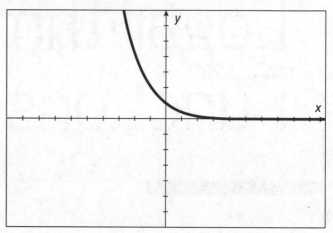

EXAMPLE

Sketch the graph of the function $f(x) = 2^{x+1}$.

The graph of $y = 2^x$ passes through the points (0, 1), (1, 2), and (2, 4) so the graph of $f(x) = 2^{x+1}$ will be shifted 1 unit to the left and will pass through (–1, 1), (0, 2), and (1, 4).

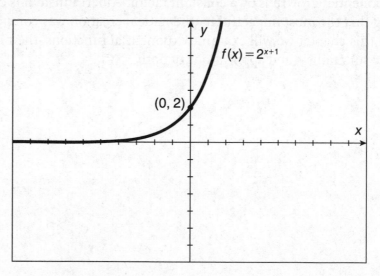

EXAMPLE

What is the range of the function $f(x) = 4(3)^{x-1} - 2$?

Since the graph is moved down 2 units, the range of the function is $y > -2$.

We use the rule if $b^x = b^y$ then $x = y$ to solve exponential equations.

EXAMPLE

Solve: $2^{2x-1} = 32$

Rewrite 32 as a power of 2, $2^{2x-1} = 2^5$. Set the exponents equal. $2x - 1 = 5$ so $x = 3$.

EXAMPLE

Solve $8^{3-x} = 128^{2x-3}$.

Since 8 and 128 are powers of 2, rewrite the equation with a base of 2, $\left(2^3\right)^{3-x} = \left(2^7\right)^{2x-3}$. Simplify each of the terms using the exponential rule $\left(b^m\right)^n = b^{mn}$, $2^{9-3x} = 2^{14x-21}$. Set the exponents equal, $9 - 3x = 14x - 21$, and solve, $30 = 17x$ so $x = \dfrac{30}{17}$.

EXAMPLE

Solve $45\left(25^{2x+5}\right) = 9\left(125^{4x-3}\right)$

45 and 25 cannot be written easily with a common base. However, if both sides of the equation are divided by 9 all remaining terms can be written as powers of 5. $45\left(25^{2x+5}\right) = 9\left(125^{4x-3}\right)$ becomes $5\left(25^{2x+5}\right) = \left(125^{4x-3}\right)$ which, in turn, becomes $5\left(5^{2(2x+5)}\right) = 5^{3(4x-3)}$. Simplify: $5^{1+2(2x+5)} = 5^{3(4x-3)}$. Set the exponents equal to each other, $1 + 2(2x + 5) = 3(4x - 3)$. Solve the equation to get $x = 2.5$.

Solve $4^x = 20$.

We know that 20 cannot be written as a power of 4 so the best option available to us is to graph the functions $f(x) = 4^x$ and $y = 20$ and determine that the point of intersection occurs when $x = 2.16$. (Realize that this is an approximate answer rounded to two decimal places.)

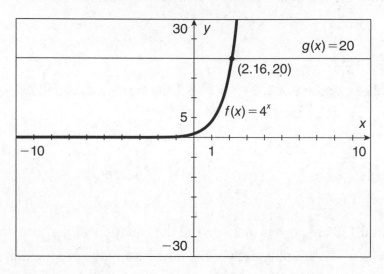

Functions of the form $P(t) = a(b^t)$ are used to model exponential growth and decay. The term a represents the initial value of P, b indicates the rate of growth or decay, and t represents the number of intervals that have occurred.

A biologist studying particular bacteria creates a colony of 1,000 of the given bacteria in a petri dish. Observation indicates that the population of the bacteria will double every 3 days. How many days are needed before the population reaches 1,024,000 bacteria?

The initial value of the population is 1,000. The rate of growth is doubling and the time frame is every 3 days. Therefore, the equation that models this data is $P(t) = 1,000 \left(2^{\frac{t}{3}} \right)$. Setting the equation equal to 1,024,000 gives $1,000 \left(2^{\frac{t}{3}} \right) = 1,024,000$. Divide by 1,000 to get $\left(2^{\frac{t}{3}} \right) = 1,024$. Given that $2^{10} = 1,024$, we set the exponents equal to determine that $t = 30$ days.

An important concept to know is that in this equation for exponential growth and decay, b represents the rate of growth. What exactly does this mean? If we look at the example of the value of a car after depreciation, then b has to be a number smaller than 1. For example, if the value of a car depreciates

10% each year, then $b = 0.9$. Why 0.9? The equation represents the value of the car from year to year. If the car loses 10% of its value, it maintains 90% of the value.

▶ Andrew purchased a new car with a value of $34,570 (price before tax, dealer preparation, etc.). One of the factors he used to purchase this particular model was that the value of the car only depreciated 8% per year. What will the value of the car be after 5 years?

▶ The car maintains 92% of its value from one year to the next. Consequently, the value of the car after 5 years will be $34,570$(0.92)^5 =$ $22,784.45.

▶ The half-life of Iodine-131 (I-131) is 192 hours. If an initial dose of 12 mCi (millicurie) is injected into the bloodstream of a patient at 8 a.m. on a Monday, when will there be less than 1 mCi remaining in the patient?

▶ The rate of decrease is $\dfrac{1}{2}$ for every 192 hours. Therefore the equation for the number of mCi of I-131 in the bloodstream is $A(t) = 12\left(\dfrac{1}{2}\right)^{\frac{t}{192}}$. To solve the inequality $A(t) < 1$, graph the function and determine the point of intersection for $y = 1$.

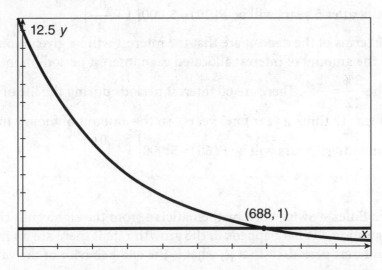

▶ The I-131 level will drop below 1 mg after 688 hours (midnight, 28 days later).

In the case of compound interest, the value of b must be larger than 1. At the end of each interest period, there will be more money in the account than there was at the beginning of each period. The tricky part to compound interest is in determining the rate of interest for each interest period. Traditional interest periods are annually, semiannually, quarterly, and monthly.

▶ Suppose that \$5,000 is deposited into an account that pays 2% compound interest. How much money will be in the account after 5 years?

- If the term of the deposit is that interest is given once per year, then the value of b will equal 100% plus the annual rate of interest, 2%. There will be 5 interest periods during the life of the problem, so the amount of money in the account will be $P(5) = 5,000(1.02)^5$.

- If the terms of the deposit are that the interest will be given semiannually, then the amount of interest allocated each interest period (6 months) will be $\dfrac{2\%}{2}$ or 1%. There are 10 interest periods during the life of the problem (twice a year for 5 years), so the amount of money in the account after 5 years will be $P(10) = 5,000(1.01)^{10}$.

- If the terms of the deposit are that the interest will be given quarterly, then the amount of interest allocated each interest period (3 months) will be $\dfrac{2\%}{4} = \dfrac{2}{4}\%$. There are 20 interest periods during the life of the problem (4 times a year for 5 years) so the amount of money in the account after 5 years will be $P(20) = 5,000\left(1 + \dfrac{.02}{4}\right)^{20}$.

- If the terms of the deposit are that the interest will be given monthly, then the amount of interest allocated each interest period (month) will be $\dfrac{2\%}{12} = \dfrac{1}{6}\%$. There are 60 interest periods during the life of the problem (12 times a year for 5 years), so the amount of money in the account after 5 years will be $P(60) = 5,000\left(1 + \dfrac{.01}{6}\right)^{60}$.

Leonhard Euler, a Swiss-born mathematician from the eighteenth century, posed the question, "What happens to the growth rate if there are an infinite number of interest periods?" That is, what is the end behavior of the value of P given $P = 1\left(1 + \dfrac{1}{n}\right)^n$ as $n \rightarrow \infty$? (It should be noted that Euler astounded most of his contemporaries with his ability to do difficult, if not tedious, calculations.) He examined the problem in which the deposit is 1 unit of currency and the rate of interest is 100%. What Euler found was that there is a bound to the amount of interest that one can get, and that number is approximately 2.7818. Years later, the mathematical community honored Euler with this accomplishment and named this number *e* after him. It turns out that Euler's number has a great deal more application to natural phenomena than just compound interest. Given that, you should include the approximate value of *e* in your repertoire as well as you know the approximate value of π is 3.14 (or better still, 3.14159).

Periods	Interest
1	2
2	2.25
100	2.7048138294215
1,000	2.7169239322359
10,000	2.7181459268252
100,000	2.718281692545
1,000,000	2.7182804693194
10,000,000	2.718281692545
100,000,000	2.7182818148676

EXAMPLE

▶ $5,000 is invested for 8 years in an account that pays 2.4% annual interest. How much money will be in the account if it is compounded (a) annually; (b) semiannually; (c) quarterly; (d) monthly; (e) continually?

▶ The amount, *A*, in the account after 8 years will be:

(a) $A = 5,000(1.024)^8 = 6,044.63$

(b) $A = 5,000\left(1 + \dfrac{0.024}{2}\right)^{8*2} = 6,051.43$

(c) $A = 5,000\left(1 + \dfrac{0.024}{4}\right)^{8*4} = 6,054.88$

(d) $A = 5,000\left(1 + \dfrac{0.024}{12}\right)^{8*12} = 6,057.19$

(e) $5,000e^{0.024(8)} = 6,058.35$

EXAMPLE

▶ Solve: $20e^{2.3x} = 100$

▶ Sketch the graph of $f(x) = 20e^{2.3x}$ and determine where the graph intersects the graph of $y = 100$.

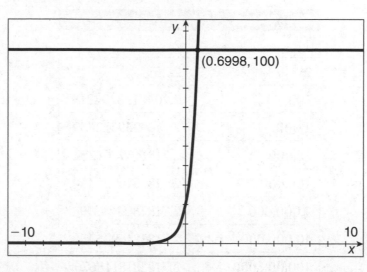

(0.6998, 100)

EXERCISE 7.1

Solve each of the equations.

1. $5(2)^{3-2x} = 160$

2. $9^{5x+3} = 81^{3x-5}$

3. $8^{4x-3} = 32^{2x+1}$

4. $6^x = 40$

5. $30\left(\dfrac{3}{4}\right)^{x-1} = 9$

6. $12e^{0.025x} = 20$

$10,000 is invested for 10 years at an annual interest rate of 4.2%. How much money is in the account if the interest is compounded:

7. Annually?

8. Quarterly?

9. Monthly?

10. Kristen's laptop computer had an initial cost of $3,500 but has depreciated 12% per year since the purchase. Kristen has had the computer for 5 years. What is the value of the computer?

Logarithmic Functions and Their Properties

As you can see in the previous section, exponential functions are 1-1 and, consequently, have inverses. (A one-to-one, 1-1, function has the property that each element in the domain has a unique element in the range and each element in the range has a unique element in the domain.) The inverse of $y = b^x$ is $x = b^y$. The problem now is to write y as a function of x. We have no algebraic process to do this. What was done to get around this problem was to define the new function as a logarithm. The equivalent of $x = b^y$ is $y = \log_b(x)$, which is read as "y equals the base b log of x." Because there are an infinite number of exponential functions, there are also an infinite number of logarithmic functions, so the base needs to be included when writing the logarithm. That is, the base needs to be identified for all logarithmic functions except two. The common logarithm, the inverse of $y = 10^x$, is written as $\log(x)$, and the natural logarithm, the inverse of $y = e^x$, is written as $\ln(x)$.

What is $\log_2(8)$? It is the exact same question as, "What is the exponent to which 2 must be raised in order to get 8?" Of course, the answer is 3. Another, and probably a more efficient, way of looking at this is when asked, "What is $\log_2(8)$?" is to say, "The answer is c, and $2^c = 8$." You will then go about the business of finding the value of c.

> It is important that you understand this very basic fact—a logarithm is an exponent!

EXAMPLE

> Evaluate $\log_4(64)$
>
> $\log_4(64) = c$ and $4^c = 64 = 4^3$, so $c = 3$.

EXAMPLE

> Evaluate $\log_4(32)$
>
> $\log_4(32) = c$ and $4^c = 32$. Both 4 and 32 are powers of 2, so rewrite $4^c = 32$ as $\left(2^2\right)^c = 2^5$. This becomes $2^{2c} = 2^5$, so $c = \dfrac{5}{2}$.

Understanding that a logarithm is just an exponent helps us to recognize three important properties of logarithms:

(1) $\log_b(mn) = \log_b(m) + \log_b(n)$

How does one determine the exponent for a product? You add the exponents.

(2) $\log_b\left(\dfrac{m}{n}\right) = \log_b(m) - \log_b(n)$

How does one determine the exponent for a quotient? You subtract the exponents.

(3) $\log_b\left(m^n\right) = n\log_b(m)$

How does one determine the exponent when an exponential statement is raised to a power? You multiply exponents.

Given $\log_b(m) = x$, $\log_b(n) = y$, and $\log_b(p) = z$. Express each of the following in terms of x, y, and z.

(a) $\log_b\left(\dfrac{m^3 n^2}{p}\right)$

(b) $\log_b\left(\dfrac{m^3}{n^2 p}\right)$

(c) $\log_b\left(\dfrac{m^3 \sqrt{p^3}}{n^2}\right)$

(d) $\left(\log_b\left(\dfrac{m^3}{n^2}\right)\right)^2$

(e) $\log_b\left(m^3 + n^2\right)$

Solutions:

(a) Use the quotient rule (2) to rewrite:

$$\log_b\left(\frac{m^3 n^2}{p}\right) \text{ as } \log_b\left(m^3 n^2\right) - \log_b(p).$$

Use the product rule (1) to change this expression to:

$$\log_b\left(m^3\right) + \log_b\left(n^2\right) - \log_b(p)$$

Use the exponential rule (3) to get:

$$3\log_b(m) + 2\log_b(n) - \log_b(p).$$

Finally, use the given information to get:

$$3x + 2y - z.$$

(b) Use the quotient rule (2) to rewrite:

$$\log_b\left(\frac{m^3}{n^2 p}\right) \text{ as } \log_b\left(m^3\right) - \log_b\left(n^2 p\right)$$

Use the product rule (1) to change this expression to:

$$\log_b\left(m^3\right) - \left(\log_b\left(n^2\right) + \log_b(p)\right)$$

Note: The extra parentheses are extremely important because the product occurs in the denominator of the fraction.

Distribute the negative and apply the exponential rule (3) to get:

$$3\log_b(m) - 2\log_b(n) - \log_b(p)$$

Substituting with the original information gives the answer:

$$3x - 2y - z$$

Mentally, compute $\dfrac{36}{2\times 3}$. Now do the problem with your calculator. Many people will get the answer 54 because they will type 36 ÷ 2 × 3 rather than 36 ÷ (2 × 3) or 36 ÷ 2 ÷ 3.

(c) Use the quotient rule (2) to rewrite:

$$\log_b\left(\frac{m^3\sqrt{p^3}}{n^2}\right) \text{ as } \log_b\left(m^3\sqrt{p^3}\right) - \log_b\left(n^2\right)$$

Use the product rule (1) to change this expression to:

$$\log_b\left(m^3\right) + \log_b\left(\sqrt{p^3}\right) - \log_b\left(n^2\right)$$

The expression $\sqrt{p^3}$ can be written as $p^{\frac{3}{2}}$.

Applying the exponential rule (3) to $\log_b\left(m^3\right) + \log_b\left(\sqrt{p^3}\right) - \log_b\left(n^2\right)$ gives:

$$3\log_b(m) + \frac{3}{2}\log_b(p) - 2\log_b(n)$$

So:

$$\log_b\left(\frac{m^3\sqrt{p^3}}{n^2}\right) = 3x + \frac{3}{2}z - 2y.$$

(d) It isn't very often that you will be asked to square a logarithm, but that is exactly what $\left(\log_b\left(\dfrac{m^3}{n^2}\right)\right)^2$ is doing. Evaluate the logarithmic expression within the parentheses first.

$$\log_b\left(\frac{m^3}{n^2}\right) = 3\log_b(m) - 2\log_b(n), \text{ which in turn equals } 3x - 2y.$$

So,

$$\left(\log_b\left(\frac{m^3}{n^2}\right)\right)^2 = \left(3x - 2y\right)^2 = 9x^2 - 12xy + 4y^2.$$

(e) This is something of a trick question but an important one nonetheless. The expression $\log_b\left(m^3 + n^2\right)$ asks for the exponent of two terms when those terms are added together. There is no rule for simplifying $b^p + b^q$, so this logarithmic expression cannot be simplified.

EXAMPLE

▶ Use the definition of a logarithm to evaluate: $\log_b(1); \log_b(b)$

$\log_b(1) = c$ becomes $b^c = 1 = b^0$ so $c = 0$. $\log_b(1) = 0$

$\log_b(b) = c$ so $b^c = b = b^1$ so $c = 1$. $\log_b(b) = 1$

These statements are true for any legitimate value of b.

If $\log_b(3) = m$ and $\log_b(5) = n$, express each of the following in terms of m and n.

(a) $\log_b(9)$

(b) $\log_b(75)$

(c) $\log_b(0.6)$

(d) $\log_5(3)$

Solutions:

(a) $\log_b(9) = \log_b(3^2) = 2\log_b(3) = 2m$

(b) $\log_b(75) = \log_b(3 \times 25) = \log_b(3) + \log_b(25) = \log_b(3) + \log_b(5^2) = m + 2n$

(c) $\log_b(0.6) = \log_b\left(\dfrac{3}{5}\right) = \log_b(3) - \log_b(5) = m - n$

(d) This one is a tad more tricky. There is no base b in this problem, and the only reference we have is base b. So, we go back to basics and let $\log_5(3) = c$ so that $5^c = 3$. Take the base b logarithm of both sides of this equation to get $\log_b(5^c) = \log_b(3)$ so that $c\log_b(5) = \log_b(3)$ and $c = \dfrac{\log_b(3)}{\log_b(5)} = \dfrac{m}{n}$.

This problem is usually referred to as the **change of base** formula: $\log_m(n) = \dfrac{\log_b(n)}{\log_b(m)}$.

Every calculator has two logarithm buttons included, **log** (the common log) and **ln** (the natural log). There are some calculators that have the feature when you press the log button. What appears is there is a box for the base of the logarithm system and a box for the argument of the function, $\log_{()}(\ )$. If your calculator has this capability you are able to evaluate $\log_3(7) = 1.77124$. If not, then you will need to apply the change of base formula and enter either $\dfrac{\log(7)}{\log(3)}$ or $\dfrac{\ln(7)}{\ln(3)}$ to get the same result. (In the case of the common logarithm, you can leave the base of the log statement blank so that, by default, the base will be 10.)

Solve $8^x = 32$

Because 8 and 32 are both powers of 2, you can rewrite the problem as $2^{3x} = 2^5$ so that $x = \dfrac{5}{3}$ or you can use your calculator to enter $\dfrac{\log(32)}{\log(8)}$ to get 1.6666667.

▶ Solve $8^x = 31$

▶ Because 8 and 31 do not share a common base, you can take the logarithm of both sides of the equation to get $\log(8^x) = \log(31)$, which then becomes $x\log(8) = \log(31)$. Divide both sides of the equation by log (8) to get $x = \dfrac{\log(31)}{\log(8)} = 1.6514$ by using your calculator.

We began this section by stating that the logarithmic function is the inverse of the exponential function. As a consequence of this relationship, the domain of the logarithmic function must the same as the range of the corresponding exponential function, namely $x > 0$, and the range of the logarithmic functions is the set of real numbers.

▶ Determine the domain of the function $g(x) = \log_b(x+5)$. (The base was left as b because the solution will be true for all values of $b > 0$ and not equal to 1.)

▶ Set the argument of the function greater than zero. $x + 5 > 0$ becomes $x > -5$.

The properties of logarithms can, at times, interfere with our ability to determine the domain of a logarithmic function.

▶ Determine the domain of the function $p(x) = \log_b\left(\dfrac{x-5}{x+2}\right)$.

▶ The domain requires that $\dfrac{x-5}{x+2} > 0$. Perform the signs analysis to determine that $x < -2$ or $x > 5$ satisfy this requirement.

▶ How do the properties of logarithms get in the way?
Rewrite the function using the properties of logarithms,

$$p(x) = \log_b\left(\dfrac{x-5}{x+2}\right) = \log_b(x-5) - \log_b(x+2).$$ The domain for $\log_b(x-5)$ is $x > 5$, while the domain for $\log_b(x+2)$ is $x > -2$. One might be led to believe that only those values greater than 5 will suffice but, as we saw, that is not so.

EXERCISE 7.2

Use this information for questions 1–4:

Given $\log_b(m) = x$, $\log_b(n) = y$, and $\log_b(p) = z$. Express each of the following in terms of x, y, z, and constants.

1. $\log_b\left(\dfrac{m^4 n^2}{\sqrt{p^5}}\right)$

2. $\log_b\left(\dfrac{m^3}{n^2 \sqrt[3]{P^2}}\right)$

3. $\log_b\left(\sqrt{\dfrac{m^3}{\sqrt[4]{p^3}}}\right)$

4. $\log_{p^2}(m^3)$

Use the following information for questions 5–8. Express each of the following in terms of x, y, z, and constants.

Given $\log_b(2) = x$, $\log_b(3) = y$, and $\log_b(5) = z$.

5. $\log_b(24)$

6. $\log_b\left(\dfrac{5}{12}\right)$

7. $\log_b(30b^2)$

8. $\log_b(180)$

9. Solve: $12^x = 217$

10. Solve: $7^{2x-1} = 74$

11. Determine the domain of the function $k(x) = \log_3\left(\dfrac{x^2 - x - 12}{x+1}\right)$

Solving Logarithmic Equations

There are a number of applications for exponential and logarithmic functions. Before we take a look at them, let's take a look at solving logarithmic equations. Basically, the goal is to get the equation to read the base b log of some expression is equal to some value and then use the definition of the logarithm to get the equation into a form that you can solve. Don't try to reread that sentence—while it is correct, it's easier to look at an example.

EXAMPLE

Solve $\log_3(x^2 - 3x - 9) = 2$

Use the definition of the logarithm to rewrite the equation as $x^2 - 3x - 9 = 3^2$.

Bring the constant to the left:

$$x^2 - 3x - 18 = 0$$

▶ Factor and solve:

$$(x-6)(x+3)=0 \text{ so } x=-3, 6$$

▶ Check both values in the original equation to show that they both satisfy the equation.

▶ Solve $\log_3(x-6)+\log_3(x+3)=2$

▶ Use the properties of logarithms to combine the two terms on the left side of the equation:

▶ $\log_3(x-6)+\log_3(x+3)=2$ becomes

$\log_3((x-6)(x+3))=\log_3(x^2-3x-18)=2$. This is the same equation we just solved, so $x=-3, 6$. When we check these values in the original problem, neither works. With $x=-3$, each of the terms $\log_3(x-6)$ and $\log_3(x+3)$ are undefined because the argument of the logarithmic function must be greater than 0. While $\log_3(x+3)$ is defined when $x=6$, $\log_3(x-6)$ is not defined. It is necessary that you check your answers in the original equations before claiming that they are solutions to a problem.

▶ Solve $4\log_3\left(\dfrac{10x+7}{x+3}\right)+5=13$

▶ To get this equation into the form where we can apply the definition of the logarithm, we'll need to move the 5 and 4 from the problem to isolate the logarithm. Subtract 5 and divide by 4 to get $\log_3\left(\dfrac{10x+7}{x+3}\right)=2$.

▶ Use the definition of the logarithm to rewrite the equation as $\dfrac{10x+7}{x+3}=3^2$.

▶ Multiply by the denominator: $10x+7=9(x+3)$. Distribute and solve: $10x+7=9x+27$ so that $x=20$.

Solve $3.2\left(7^{2x-3}\right)-8=1{,}343$

Add 8 and divide by 3.2 to get $7^{2x-3}=\dfrac{1{,}351}{3.2}$. Don't bother finding the quotient because we need to take the logarithm of both sides of the equation and we can let the calculator hold onto all the decimal places. Taking the log of both sides of the equation gives $(2x-3)\log(7)=\log\left(\dfrac{1{,}351}{3.2}\right)$. (We could have used the natural logarithm here as well.) We'll solve for x and will not use the calculator until the last step. Solve $2x-3=\dfrac{\log\left(\dfrac{1{,}351}{3.2}\right)}{\log(7)}$ to get $x=\dfrac{\dfrac{\log\left(\dfrac{1{,}351}{3.2}\right)}{\log(7)}+3}{2}$. The calculator strokes for getting this answer require the use of parentheses: $\left(\log(1{,}351\,/\,3.2)\,/\,\log(7)+3\right)/2=3.05337$.

If the thought of entering that many keystrokes concerns you, you have options. You can use the calculator's memory function. The key piece is to allow the calculator to hold onto all the decimal places. Don't divide 1,351 by 3.2, approximate the quotient as 422.2 and continue from there. The calculations that follow may exacerbate the difference between the exact and approximate values and lead to an incorrect answer.

Solve the equation $\log_2(x+3)=x-3$

There is no way to solve these equations algebraically. Set up two functions, $f(x)=\log_2(x+3)$ and $g(x)=x-3$, and graph these on the same set of axes.

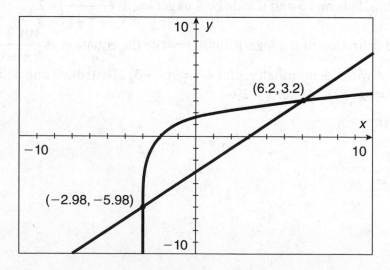

EXAMPLE

The half-life of Iodine-131 (I-131) is 8 days. If an initial dose of 12 mCi (millicuries) is injected into the bloodstream of a patient at 8 a.m. on a Monday, when will there be less than 1 mCi remaining in the patient?

We saw this problem earlier in the chapter, and the solution used was graphical. We will take an algebraic approach this time. We'll solve for the time when there is exactly 1 mCi and know that from the next moment on, the amount of I-131 in the bloodstream will be less than 1. The function describing the amount of I-131 in the bloodstream is $A(t) = 12\left(\dfrac{1}{2}\right)^{\frac{t}{192}}$.

Because $\dfrac{1}{2} = 2^{-1}$, let's write the function as:

$$A(t) = 12(2)^{\frac{-t}{192}}$$

Set $A(t) = 1$ and solve:

$$12(2)^{\frac{-t}{192}} = 1.$$

Divide by 12:

$$(2)^{\frac{-t}{192}} = \frac{1}{12}.$$

Take the logarithm of both sides of the equation:

$$\frac{-t}{192}\log(2) = \log\left(\frac{1}{12}\right) = -\log(12)$$

Multiply both sides of the equation by $\dfrac{-192}{\log(2)}$:

$$t = \frac{192\log(12)}{\log(2)}$$

Use your calculator to determine that t is approximately 688.3.

EXAMPLE

The population of an urban area is currently 6.74 million people, and a mathematical model for the future population is given by $P(t) = 6.74e^{0.032t}$, where t represents the number of years after 2018. (That is, $t = 0$ represents the beginning of 2018.) What is the projected population for 2025? Under this model, in what year will the population reach 10 million?

2025 is 7 years after 2018, so $P(7) = 6.74e^{0.032*7} = 8.43$ million. When will $P(t) = 10$ million? Divide $6.74e^{0.032t} = 10$ by 6.74 to get $e^{0.032t} = \dfrac{10}{6.74}$. Take the natural logarithm of both sides of the equation: $0.032t \ln(e) = \ln\left(\dfrac{10}{6.74}\right)$. Recall that $\log_b(b) = 1$ and that ln means the base e log, $\ln(e) = 1$. Therefore, $0.032t \ln(e) = \ln\left(\dfrac{10}{6.74}\right)$ becomes $0.032t = \ln\left(\dfrac{10}{6.74}\right)$ which then becomes $t = \dfrac{\ln\left(\dfrac{10}{6.74}\right)}{0.032}$. Use your calculator to find that $t = 12.3$. The population of the urban area will reach 10 million people during the year 2030.

EXERCISE 7.3

Solve each of the equations. Express approximate answers to 2 decimal places.

1. $x = \log_5(210)$

2. $5^{3x-7} = 468$

3. $17(2.3)^x - 9 = 861$

4. $19.3\,e^{-0.24x} + 8.3 = 12.5$

5. $\log_2(x^2 - 8x - 1) = 6$

6. $\log_4\left(\dfrac{x^2 + 42x - 8}{x - 2}\right) = 3$

7. $\log_4(x + 3) = x^2 - 2$

The amount of Iodine-131 in the bloodstream can be modeled by the equation $A(t) = 10e^{-.00361t}$ where t represents the number of hours after 10 mCi of I-131 are introduced into the bloodstream.

8. How many millicuries of I-131 are in the bloodstream 24 hours after the I-131 has entered the bloodstream?

9. How many hours are required before there are only 5 mCi of I-131 in the bloodstream?

CHAPTER 8

Sequences and Series

Much of the study of mathematics is looking at patterns. Some patterns are very basic;

> 1, 2, 3, 4, 5, . . .
> a, b, c, d, e, . . .

Some are a bit more complicated:

> 1, 4, 7, 10, 13, . . .
> 3, 6, 12, 24, 48, . . .

Some are just tricky:

> O, T, T, F, F, S, S, E, N, T, . . .

Can you identify the next three terms in this sequence?

> 1, 1, 2, 3, 5, 8, 13, 21, 34, . . .

While this pattern looks as tricky as the last example, it has applications in a variety of studies. We will look more into this pattern later in the chapter.

Some patterns fit each of these styles and are nothing more than curiosities, while others lead to some interesting mathematics. In this chapter, we will examine both sequences and series from three perspectives—by list, by recursion, and by formula. (Did you determine that the next three terms in the pattern O, T, T, F, F, S, S, E, N, T are E, T, T? The pattern is the first letter in the words One, Two, Three, Four, Five, . . . and then Eleven, Twelve, Thirteen.)

Summation Notation

Before we actually get to the topics of sequences and series, there are a couple of items to examine. The first of these is notation for repeated addition. The uppercase Greek letter sigma, Σ, is used to represent addition. The notation for this process is $\sum_{begin}^{end} rule$, where we list the beginning and ending values for adding and "rule" represents a pattern generator. The notation is read as "the summation from the beginning value to the ending value of the rule." For example, $\sum_{1}^{6} 5$ is the summation from 1 to 6 of 5 and is equal to $5 + 5 + 5 + 5 + 5 + 5 = 30$. That is, we are told to add the number 5 six times. As a consequence, if c is some constant, then $\sum_{1}^{n} c = nc$ because the number c is being added n times.

When the rule is given by some formula, things get much more interesting. A simple example is $\sum_{n=1}^{5} 5n + 1$. We identify the variable in the rule with the beginning element as well as in the rule itself. (When the beginning element does not name the variable, it is assumed to be the same as that in the rule. If the rule contains multiple symbols, such as $5n + b$, we need to identify the variable and treat the other character as a constant.)

$$\sum_{n=1}^{5} 5n + 1 = \left(5(1)+1\right) + \left(5(2)+1\right) + \left(5(3)+1\right) + \left(5(4)+1\right) + \left(5(5)+1\right) = 6 + 11 + 16 + 21 + 26 = 80$$

The beginning value of the summation does not have to be 1.

EXAMPLE

> What is the value of $\sum_{n=2}^{4} 3n^5$?
>
> $$\sum_{n=2}^{4} 3n^5 = 3(2)^5 + 3(3)^5 + 3(4)^5 = 96 + 729 + 3{,}072 = 3{,}897.$$

There are two other important properties of the summation process:

(1) $\sum_{1}^{n}\left(a_i + b_i\right) = \sum_{1}^{n}\left(a_i\right) + \sum_{1}^{n}\left(b_i\right)$

(2) If c is a constant, $\sum_{i}^{n}\left(ca_i\right) = c\sum_{i}^{n}\left(a_i\right)$.

If $\displaystyle\sum_1^{10} a_i = 120$ and $\displaystyle\sum_1^{10} b_i = 150$, compute $\displaystyle\sum_1^{10}\left(a_i + 3b_i\right)$.

$$\sum_1^{10}\left(a_i + 3b_i\right) = \sum_1^{10}\left(a_i\right) + \sum_1^{10}\left(3b_i\right) = \sum_1^{10}\left(a_i\right) + 3\sum_1^{10}\left(b_i\right) = 120 + 3(150) = 570.$$

EXERCISE 8.1

Evaluate each of the following.

1. $\displaystyle\sum_1^{10}(n)$

2. $\displaystyle\sum_1^{10}\left(n^2\right)$

3. $\displaystyle\sum_1^{10}\left(n^2 - 2n\right)$

4. $\displaystyle\sum_1^{10}\left(2^n\right)$

5. $\displaystyle\sum_1^{10}\left(2^n + n^2 - 2n\right)$

Sequences

A sequence is a listing of terms in a pattern. The rule that defines the pattern may be any algebraic expression, but the domain of the rule is the set of counting numbers (which is the same as the set of positive integers).

List the first five elements of the sequence determined by the function $f(n) = 5n$.

$f(1) = 5, f(2) = 10, \ldots f(5) = 25$ so the first five terms of the sequence are 5, 10, 15, 20, 25.

List the first 5 elements of the sequence determined by the function $g(n) = n^3$.

$g(1) = 1, g(2) = 8, g(3) = 27, g(4) = 64$, and $g(5) = 125$ so the first five terms of the sequence are 1, 8, 27, 64, and 125.

Consider the following problem: Alice asks Maria to pick a number, and Maria responds, "5." Alice then asks Maria to pick a second number, to which Maria responds, "7." Alice directs Maria to take the first number, add the second number, and to continue to add the second number to the previous sum. The result of this exercise yields the results:

5, 12, 19, 26, 33, . . .

You should recognize that these numbers with their constant difference between them would lie on a line. In this case, the equation of that line is $y = 7x - 2$. If asked what the tenth number in this pattern would be, Maria could substitute 10 for x and get 68 as the result.

The directions given to Maria by Alice lead to another approach. Alice asks Maria to pick a number, and Maria responds, "5." The first term, $a_1 = 5$. Maria is then told "take the first number, add the second number" gives $a_2 = a_1 + 7 = 12$. Then Maria is directed "to continue to add the second number to the previous sum" indicating that $a_3 = a_2 + 7$, $a_4 = a_3 + 7$, $a_5 = a_4 + 7$ and so on. After the first term is established, all other terms follow the rule $a_1 = 5, a_n = a_{n-1} + 7$. This process is called **recursion**.

EXAMPLE

▸ Write the first five terms of the sequence defined recursively as $a_1 = 3$ and $a_n = 4a_{n-1}$.

▸ Interpret what the recursive formula is saying: "The first term is 3, and every term after that is 4 times the previous value." The first five terms in the sequence are 3, 12, 48, 192, 768.

Can this sequence of numbers be written with an algebraic rule? Let's look at the results, written in a different manner. The second term is 3×4. The third term is $3 \times 4 \times 4 = 3 \times 4^2$. The fourth term is $3 \times 4^2 \times 4 = 3 \times 4^3$, and the fifth term is $3 \times 4^3 \times 4 = 3 \times 4^4$. The rule for this sequence of numbers is $k(n) = 3 \times 4^{n-1}$.

EXAMPLE

▸ Write the first five terms of the sequence defined recursively as $a_1 = 2$ and $a_n = 3a_{n-1} + 2$.

▸ The first of the sequence is 2. The second term is $3(2) + 2 = 8$. The third term is $3(8) + 2 = 26$. The fourth term is $3(26) + 2 = 80$, and the fifth term is $3(80) + 2 = 242$. To recap, the five terms are 2, 8, 26, 80, 242.

Can this sequence be written with an algebraic rule? The second number is $3(2) + 2 = 8$. The third number is

$$3(8) + 2 = 3(3(2) + 2) + 2 = 3^2(2) + 3(2) + 2 = 2(3^2 + 3 + 1) = 26.$$

The fourth number is $3(26) + 2 = 3(26) + 2 = 3(3^2(2) + 3(2) + 2) + 2 =$

$3^3(2) + 3^2(2) + 3(2) + 2 = 2(3^3 + 3^2 + 3 + 1) = 80$. While there is a clear pattern emerging, it is not one that leads to an algebraic rule.

Sequences identified in a recursive definition can have more than one constant to initiate the recursion. For example, the sequence defined by $a_1 = 1, a_2 = 1, a_n = a_{n-2} + a_{n-1}$ is the famous Fibonacci sequence: 1, 1, 2, 3, 5, 8, 13, . . .

Every term after the second term is found by adding the two previous terms together. This sequence has plenty of applications in the world of biology from the breeding of adult rabbits to the number of seeds in a sunflower.

EXAMPLE

> Write the first eight terms of the sequence defined as $a_1 = 2, a_2 = 5, a_n = a_{n-1} - a_{n-2}$.

> The first two terms are clearly 2 and 5. The third term $a_3 = a_{3-1} - a_{3-2} = a_2 - a_1 = 5 - 2 = 3, a_4 = 3 - 5 = -2, a_5 = -2 - 3 = -5; a_6 = -5 - (-2) = -3; a_7 = -3 - (-5) = 2; a_8 = 2 - (-3) = 5$. The first eight terms of the sequence are 2, 5, 3, -2, -5, -2, 3, 5.

EXERCISE 8.2

For exercises 1–4, determine the first five terms for the sequence defined.

1. $f(n) = 3n + 4$

2. $k(n) = 7(3^{n+1})$

3. $h(n) = 2,400 - 120n$

4. $p(n) = 400,000\left(\dfrac{1}{2}\right)^n$

For exercises 5–8, determine the first six terms of the sequence defined.

5. $a_1 = 9, a_n = a_{n-1} + 4$

6. $a_1 = 9, a_n = 3a_{n-1} + 4$

7. $a_1 = 9, a_n = 3(a_{n-1} - 4)$

8. $a_1 = 2, a_2 = 7; a_n = 3a_{n-1} + a_{n-2}$

9. Rewrite the sequence 3, 10, 17, 24, 31, 38, . . . as a function $f(n)$.

10. Rewrite the sequence 3, 18, 108, 648, 3,888, 12,528, . . . in recursive form.

Arithmetic Sequences

One definition of an arithmetic sequence is one that is the output of a linear function. (Remember that the domain of a sequence is the set of counting numbers.) A second definition for an arithmetic sequence is one in which consecutive terms have a constant difference. You "studied" arithmetic sequences in early elementary schools when you were taught to skip count. This definition is the recursive formula for the sequence. The sequence generated by Alice and Maria in the last section, $a_1 = 5, a_n = a_{n-1} + 7$, is arithmetic.

EXAMPLE

▶ The seating in the lower level of the Ming-Sun Theater is divided into three sections—stage left, center, and stage right. There are 21 seats in the first row of the center section. For each row behind the first, there are two more seats in that row than the row ahead of it. How many seats are in the twenty-fifth row?

▶ The formula defining the number of seats in a row of the center section is $a_1 = 21, a_n = a_{n-1} + 2$. The trouble with recursive definitions is that in order to determine the twenty-fifth term, you must determine the previous 24 terms. The functional definition, $f(n) = 2n + 19$ will be much easier to work with. Do you understand how the 2 and 19 are determined? The constant difference between the numbers of seats in consecutive rows, the slope, is 2. The number of seats in the first row is 21, so the "y-intercept" or "zeroth term" must be 2 less than this. Consequently, the number of seats in the twenty-fifth row, $f(25)$, is 69.

EXAMPLE

▶ Find the one hundredth term of the sequence defined by $a_1 = 17, a_n = a_{n-1} + 3$.

▶ As was stated in the previous paragraph, finding a specific term from a recursively defined sequence can be difficult. We'll translate this definition to function notation. What is the constant difference between terms? The answer is 3 from $a_n = a_{n-1} + 3$. The first term in the sequence is 17. The "term that would precede this," if such a term existed, would be 14. Therefore, the function defining this sequence is $f(n) = 3n + 14$. The one hundredth term of the sequence is $3(100) + 14 = 314$.

EXAMPLE

▶ Find the fortieth term of the sequence defined by $a_1 = 170, a_n = a_{n-1} - 4$.

▶ The slope for the functional definition is –4 (the terms are decreasing in value) so the zeroth term must be 4 more than 170, 174. The functional definition is $f(n) = 174 - 4n$ making $f(40) = 14$.

EXAMPLE

▶ The twelfth term of an arithmetic sequence is 21, and the thirty-first term of the sequence is 78. What is the fiftieth term of the sequence?

▶ Using the function definition $f(n) = dn + b$ (d representing the common difference), $f(12) = 21$ and $f(31) = 78$, we get the system of equations:

$$31d + b = 78$$

$$12d + b = 21$$

▶ Subtract the two equations to get $19d = 57$, so $d = 3$. Substitute 3 for d in one of the equations to determine $b = -15$. Therefore, $f(50) = 3(50) - 15 = 135$.

▶ A third notation for writing arithmetic sequences is similar to the functional notation. Take the function $f(n) = 3n - 15$. The term –15 is not really an element of the sequence because it is associated with $n = 0$ and 0 is not a counting number. What is done instead is to alter the formula. The sequence $f(n) = 3n - 15$ can be written as $f(n) = 3n - 3 - 12 = 3(n-1) - 12$. Written this way, the first term of the sequence is used in the definition. Finally, rather than write $f(n) = 3(n-1) - 12$, the notation $a_n = 3(n-1) - 12$ is used.

EXERCISE 8.3

1. The number of seats in the first row of the stage left section of the Ming-Sun Theater is 9. As with the center section, the number of seats in each succeeding row is 2 more than the row in front of it. How many seats are in the twenty-fifth row of the stage left section?

2. The nineteenth term in an arithmetic sequence is 243, and the eleventh term is 147. What is the value of the eighty-sixth term?

3. Find the forty-third term of the sequence defined by $a_1 = 1,916, a_n = a_{n-1} - 163$.

4. Find the thirty-fourth term of the sequence defined by $a_1 = 146, a_n = a_{n-1} + 36$.

Arithmetic Series

A series is the sum of the terms in a sequence, so an arithmetic series is the sum of the terms in an arithmetic sequence. Let S represent the sum: $S = a_1 + a_2 + a_3 + \ldots + a_{n-2} + a_{n-1} + a_n$. Write the sum again, except write the terms from last term to first term: $S = a_n + a_{n-1} + a_{n-2} + \ldots + a_3 + a_2 + a_1$. When you add these equations together, you get $2S = (a_1 + a_n) + (a_1 + a_n) + (a_1 + a_n) + \ldots + (a_1 + a_n) + (a_1 + a_n) + (a_1 + a_n)$. The right-hand side of this equation comprises n terms, each of which is the sum of the first and last term. Writing the right-hand side as $n(a_1 + a_n)$, the equation becomes $2S = n(a_1 + a_n)$, so the sum of the first n terms of the arithmetic series, S, is equal to one-half the number of terms multiplied by the sum of the first and last terms. That is, $S = \dfrac{n}{2}(a_1 + a_n)$.

EXAMPLE

> Find the sum of the terms in the sequence 1, 2, 3, 4, 5, 6, 7, 8, 9, 10.

> The first term is 1 and the last of the 10 terms is 10, so the sum is $\dfrac{10}{2}(1 + 10) = 55$.

EXAMPLE

> Find the sum of the first n counting numbers.

$$1 + 2 + 3 + \ldots + n = \frac{n}{2}(1 + n) = \frac{n(n+1)}{2}$$

EXAMPLE

> What is the sum of the first n even positive integers?

> The sum of the first n positive integers is:

$$2 + 4 + 6 + \ldots + 2n = 2(1 + 2 + 3 + \ldots + n) = 2\left(\frac{n(n+1)}{2}\right) = n(n+1).$$

EXAMPLE

> Find the sum of the first 50 terms in the sequence defined by $f(n) = 12n - 3$.

> The number of terms is 50, the first term is $f(1) = 12 - 3 = 9$, and the fiftieth term is $f(50) = 12(50) - 3 = 597$. The sum of the first 50 terms is $S = \dfrac{50}{2}(9 + 597) = 15{,}150$. The sum of the first n terms can be written as S_n.

EXAMPLE

Find the sum $23 + 34 + 45 + 56 + \ldots + 232$.

The difference between consecutive terms is 11, so we know the series is arithmetic. We know the values of the first and last terms. We need to determine the number of terms in the series before we can compute the sum. The rule defining the series can be written as $f(n) = 11(n - 1) + 23$ (or $f(n) = 11n + 12$). Solve $f(n) = 232$ to determine that $n = 20$. Therefore,

$$S_{20} = \frac{20}{2}(23 + 232) = 2{,}550.$$

EXAMPLE

Compute $\displaystyle\sum_{n=1}^{25} 4n + 13$

We are asked to find the sum of the first 25 terms of the arithmetic sequence defined by:

$$f(n) = 4n + 13. \; S_{25} = \frac{25}{2}(17 + 113) = 1{,}625$$

EXAMPLE

Compute $\displaystyle\sum_{n=11}^{25} 4n + 13$

This problem asks that we solve the sum of the series from the eleventh term to the twenty-fifth term. We just computed the sum of the first 25 terms. If we compute the sum of the first 10 terms we can subtract the results to get the answer:

$$\sum_{n=11}^{25} 4n + 13 = \sum_{n=1}^{25} 4n + 13 - \sum_{n=1}^{10} 4n + 13 = 1{,}625 - 350 = 1{,}275$$

EXERCISE 8.4

Compute each of the following.

1. The sum of the first 80 terms of the sequence determined by $f(n) = 7n + 4$.

2. $\displaystyle\sum_{n=1}^{60} 19n - 8$

3. $\displaystyle\sum_{n=16}^{50} 5n + 17$

Questions 4–6 refer to the lower level seating of the Ming-Sun Theater from the previous section. Assume each section has 25 rows.

4. Determine the number of seats in the stage left section of the lower level of the Ming-Sun Theater.

5. Determine the number of seats in the center section of the lower level of the Ming-Sun Theater.

6. The first row of seats in the stage right section of the theater contains 11 seats, and each subsequent row contains 2 more seats than the previous row. Determine the total number of seats on the lower level of the theater.

Geometric Sequences

A sequence is called geometric when the ratio of consecutive terms is a constant. From a functional perspective, a geometric sequence is the range of an exponential function (again, with the domain being the set of positive integers). The rules $f(n) = 2(3^n)$ and $a_1 = 6, a_n = 3a_{n-1}$ generate the same sequence.

EXAMPLE

List the first five terms of the sequence $a_1 = 10{,}000, a_n = 1.01a_{n-1}$.

The terms are 10,000, 10,000(1.01), 10,000(1.01)(1.01) = 10,000(1.01)², 10,000(1.01)³, 10,000(1.01)⁴. When computed, these numbers are 10,000, 10,100, 10,201, 10,303.01, and 10,406.04.

One of the applications of geometric sequences is compound interest. The numbers represented in this last example show how $10,000 will grow when it is in an account paying 1% interest each interest period (1% compounded annually or 4% compounded quarterly, for example).

EXAMPLE

Find the tenth term of the sequence defined by $a_1 = 35{,}000$, $a_n = 0.85a_{n-1}$.

As is the case with arithmetic sequences, one has to find all the terms in a sequence before getting to the one you are looking for when the sequence is defined recursively. The first term of this sequence is 35,000 and the second term is 35,000(0.85). The third, fourth, and fifth terms are, respectively, $35{,}000(0.85)^2$, $35{,}000(0.85)^3$, and $35{,}000(0.85)^4$. Do you see how the exponent on the constant factor is 1 less than the number of the term? That is $f(n) = a_n = 35{,}000(0.85)^{n-1}$. The tenth term of the sequence is 8,106.593 (answer rounded to three decimal places).

> Sequences with a common difference, arithmetic sequences, are defined by a linear function. Sequences with a common ratio, geometric sequences, are defined by an exponential function.

EXAMPLE

Find the twelfth term of a geometric sequence with all positive terms if the third term is 72 and the seventh term is 5,832.

If we write the generating equation for the sequence as $f(n) = ar^{n-1}$ where a is the initial term and r is the common factor, the third term is $72 = ar^2$ and the seventh term is $5{,}832 = ar^6$. We can eliminate the value of a by rewriting ar^6 as $(ar^2)r^4$ and substituting 72 for $a(r)^2$. Solve the equation $5{,}832 = 72r^4$. Divide by 72 to get $81 = r^4$, and take the fourth root to get $r = 3$. Substitute for r in $72 = ar^2$ getting $72 = a(9)$ so $a = 8$. The twelfth term of the sequence is $8(3)^{11} = 1{,}417{,}176$.

EXAMPLE

How many terms are required before a term of the geometric sequence $f(n) = 7(2.3)^{n-1}$ exceeds 1 million?

Solve the equation $7(2.3)^{n-1} = 1{,}000{,}000$. Divide by 7:
$(2.3)^{n-1} = \dfrac{1{,}000{,}000}{7}$. We'll need to use logarithms to solve this problem.

$\log(2.3)^{n-1} = \log\left(\dfrac{1{,}000{,}000}{7}\right)$ becomes $n - 1 = \dfrac{\log\left(\dfrac{1{,}000{,}000}{7}\right)}{\log(2.3)}$ or

$n = 1 + \dfrac{\log\left(\dfrac{1{,}000{,}000}{7}\right)}{\log(2.3)}$. Enter the expression on the right-hand side of the equation into your calculator to get $n = 15.250775$. The fifteenth term will not be larger than 1 million, but the sixteenth term will be.

EXERCISE 8.5

1. Find the thirteenth term of the sequence $f(n) = 10{,}000(1.005)^{n-1}$. (Answer to the nearest hundredth.)

2. The fourth term in a geometric sequence is 48 and the eighth term is 768. Find the fifteenth term.

3. How many terms are required before a term of the geometric sequence $f(n) = 75(0.6)^{n-1}$ is less than 1 ten-thousandth?

4. A standard sheet of loose-leaf paper is 0.1 mm thick. If we fold this paper once, the folded paper would be 0.2 mm thick. Suppose this piece of paper is folded in half a number of times. (Yes, this requires that we are able to fold something this small.) How thick would the paper be after the fiftieth fold?

Geometric Series

A geometric series is the sum of the terms of a geometric sequence. The sum of the first n terms, S_n, is:

$$S_n = a + ar + ar^2 + ar^3 + ar^4 + \ldots + ar^{n-2} + ar^{n-1}$$

Unlike what was done to find the formula for the arithmetic series, we will multiply both sides of the equation by r and then subtract the two equations from each other.

$$S_n = a + ar + ar^2 + ar^3 + ar^4 + \ldots + ar^{n-2} + ar^{n-1}$$
$$rS_n = ar + ar^2 + ar^3 + ar^4 + ar^5 + \ldots + ar^{n-1} + ar^n$$

Notice that all the terms, with the exception of the first term of the first equation and the last term of the second equation, drop from the problem because of the subtraction. The result of the subtraction is:

$$S_n - rS_n = a - ar^n$$

$$S_n(1 - r) = a(1 - r^n)$$

$$S_n = \frac{a(1 - r^n)}{1 - r}$$

Clearly, r cannot equal 1 as that would cause the denominator to equal zero. However, that is consistent with the conditions stated for the exponential function $f(x) = b^x$ where it was said that $b > 0$ and b cannot equal 1.

EXAMPLE

Find the sum $1 + 2 + 4 + 8 + 16 + 32 + 64 + 128 + 256 + 512$

There are 10 terms in the summation, so $S_{10} = \dfrac{1\left(1 - 2^{10}\right)}{1 - 2} = 1{,}023.$

EXAMPLE

Find the sum $1 + 2 + 4 + 8 + 16 + 32 + 64 + 128 + 256 + 512 + \ldots + 2^{n-1}.$

There are n terms in the series so $S_n = \dfrac{1\left(1 - 2^n\right)}{1 - 2} = 2^n - 1.$

EXAMPLE

Compute $\displaystyle\sum_{n=1}^{15} 3(1.2)^{n-1}$

There are 15 terms in the series with $a = 3$ and $r = 1.2$. Therefore,

$$S_{15} = \dfrac{3\left(1 - (1.2)^{15}\right)}{1 - 1.2} = 216.105 \text{ (rounded to 3 decimal places)}.$$

EXAMPLE

Compute the sum $4 + 12 + 36 + 108 + 324 + \ldots + 708{,}588.$

The first term is 4 and the common ratio is 3. Solve the equation $708{,}588 = 4(3)^{n-1}$ to determine the number of terms in the series. Dividing by 4 gives

$3^{n-1} = 177{,}147$. Use logarithms to solve for n. $n = 1 + \dfrac{\log(177{,}147)}{\log(3)} = 12$. The

sum of these terms is $S_{12} = \dfrac{4\left(1 - 3^{12}\right)}{1 - 3} = 1{,}062{,}880.$

Alternate Series

Unlike exponential functions when the base must be a positive number (other than 1), the common ratio of a geometric sequence and geometric series can be negative. The signs of the terms in the sequence will alternate signs.

Compute the sums of the first 10 terms for the series defined by $f(n)=3(2)^{n-1}$ and $g(n)=3(-2)^{n-1}$.

The series defined by $f(n)$ is $3 + 6 + 12 + 24 + 48 + 96 + 192 + 384 + 768 + 1{,}536$ and its sum is $S_{10}=\dfrac{3\left(1-2^{10}\right)}{1-2}=3{,}069$, while the series defined by $g(n)$ is $3 - 6 + 12 - 24 + 48 - 96 + 192 - 384 + 768 - 1{,}536$ and its sum is $S_{10}=\dfrac{3\left(1-(-2)^{10}\right)}{1-(-2)}=1{,}023$.

Determine the sum $1+\dfrac{1}{2}+\dfrac{1}{2^2}+\dfrac{1}{2^3}+\dfrac{1}{2^4}+\ldots+\dfrac{1}{2^{n-2}}+\dfrac{1}{2^{n-1}}$

The initial term is 1, the common ratio is $\dfrac{1}{2}$, and the number of terms is n. The sum is:

$$S_n=\frac{1\left(1-\dfrac{1}{2^n}\right)}{1-\dfrac{1}{2}}=2\left(1-\frac{1}{2^n}\right)=2-\frac{1}{2^{n-1}}$$

Given the infinite series $1+\dfrac{1}{2}+\dfrac{1}{2^2}+\dfrac{1}{2^3}+\dfrac{1}{2^4}+\ldots$, what is the sum of this infinite number of terms? The formula gives the equation $S_\infty=\dfrac{1\left(1-\dfrac{1}{2^\infty}\right)}{1-\dfrac{1}{2}}$.

However, we cannot take 2 and raise it to an infinite power. But we can talk about patterns and behavior. As the value of the exponent grows very large, the value of 2^n grows infinitely large and the value of $\dfrac{1}{2^n}$ gets close to 0. Therefore, the sum of this infinite series grows closer to 2.

$$S_\infty=\frac{1\left(1-\dfrac{1}{2^\infty}\right)}{1-\dfrac{1}{2}}=2\left(1-\frac{1}{2^\infty}\right)=2(1-0)=2.$$

Infinite Geometric Series

So long as $|r| < 1$, the value of r^n will go to zero as n goes to infinity. The sum of an infinite series is $S = \dfrac{a}{1-r}$.

> Compute: $3{,}600 - 1{,}800 + 900 - 450 + 225 - \ldots$
>
> Solution: After 3,600, consecutive terms are found by multiplying the previous term by $\dfrac{-1}{2}$. Therefore, the sum of this infinite geometric series
>
> is $\dfrac{3{,}600}{1 - \dfrac{-1}{2}} = \dfrac{3{,}600}{\dfrac{3}{2}} = 2{,}400$.

EXERCISE 8.6

Find the indicated sums.

1. $\displaystyle\sum_{n=1}^{12} 2\left(3^{n-1}\right)$

2. $\displaystyle\sum_{n=1}^{12} 2\left(\dfrac{1}{3}\right)^{n-1}$

3. $S_n = 8 + 12 + 18 + 27 + \dfrac{81}{2} + \dfrac{243}{4} + \ldots + \dfrac{19{,}683}{64}$

4. $\displaystyle\sum_{n=1}^{\infty} 2\left(\dfrac{1}{3}\right)^{n-1}$

5. $\displaystyle\sum_{n=1}^{\infty} 2\left(\dfrac{-1}{3}\right)^{n-1}$

6. $S_{49} = 1{,}000 + 1{,}000(1.05) + 1{,}000(1.05)^2 + 1{,}000(1.05)^3 + 1{,}000(1.05)^4 + \ldots + 1{,}000(1.05)^{47} + 1{,}000(1.05)^{48}$

7. Explain how the summation in problem 6 determines the amount of money in an account if a person deposits \$1,000 into an account that pays 5% compounded annually each year on her birthday.

CHAPTER 9

Trigonometry

There are a number of examples of periodic phenomena that can be described or modeled with trigonometric functions. The height of a car on a Ferris wheel, the voltage flowing with an alternating current through a circuit, and low and high tides are just a few of the topics that lend themselves to modeling with trigonometric functions. A problem that arises though is that these phenomena are dependent on time and not on the measure of an angle. To accommodate this, a system is used that is dependent only on unitless measurements. Before you take exception to this notion, consider that you know at least two ways of measuring temperature (Fahrenheit and Celsius; three measurements if you are studying chemistry and have used the Kelvin system) and two ways of measuring lengths and volume (metric and standard).

The Unit Circle—the First Quadrant

The original study of trigonometry was a strictly with right triangles, so the conversation was limited to acute angles. When we extend the study to the coordinate plane, we allow for all possible angles to be included, including negative angles. The basis for this extension reaches back to the days when the Moors occupied Spain. When the Moors were defeated and left Spain, scholars were able to examine their libraries. One of the items the scholars found was this diagram.

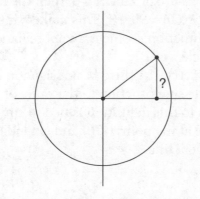

The word written where the question mark is was not a word that any of the scholars of the day understood. Given that, they translated it to the word that was closest to the Arabic word they understood. In English, that word was *pocket*. In Latin, the language in which all scholarly work in Europe was done, the word was *sinus*. It turns out the real meaning of the Arabic word was "half-chord" because that segment was half of a chord of the circle. That's the story of the sine function.

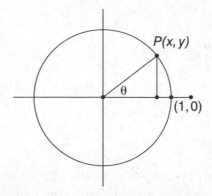

The circle with radius equal to 1 and whose center is at the origin is called the unit circle. Let P be a point on the unit circle and in the first quadrant, and let theta, θ, be the angle made between the radius containing P and the positive x-axis. The sine of theta, $\sin(\theta)$, is the y-coordinate of P, and the cosine of theta, $\cos(\theta)$, is the x-coordinate of P.

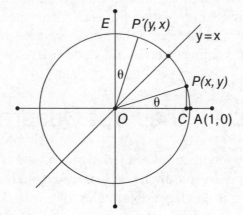

One of the transformations you learned in geometry is reflection across a line. In particular, if the line is the line with equation $y = x$, the image of point $P(x, y)$ is $P'(y, x)$. If the measure of $\angle POA = \theta$, then the measure of $\angle P'OE = \theta$ making the measure of $\angle P'OA = 90 - \theta$. This makes $\sin(\theta) = \cos(90 - \theta)$ and $\cos(\theta) = \sin(90 - \theta)$. The function cosine is an abbreviation of the function "sine of the complementary angle."

As you studied in geometry, the third basic trigonometric ratio, tangent, is the ratio of the value of the sine function to the value of the cosine function. A truly geometric approach to this definition is tied to similar triangles. Construct a tangent line to the circle at the point (1, 0) and extend the radius through P until it intersects the tangent line.

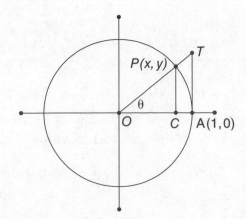

With right angles at C and A, $\triangle OCP \sim \triangle OAT$. Corresponding sides of similar triangles are in proportion so $\dfrac{TA}{OA} = \dfrac{PC}{OC}$. TA is the length of the segment on the tangent line formed by the triangles, $OA = 1$ because it is the radius of the unit circle, PC is the value of the sine function, and OC is the value of the cosine function. Consequently, $\tan(\theta) = \dfrac{\sin(\theta)}{\cos(\theta)}$.

We also have the proportion $\dfrac{OT}{OA} = \dfrac{OP}{OC}$. OT is the length of the secant segment from the center of the circle to the point of intersection with the tangent line. This gives the definition of the trigonometric function $\sec(\theta)$, $\sec(\theta) = \dfrac{1}{\cos(\theta)}$.

Extend the radius through P and draw the tangent line to the circle at the point $(0, 1)$. Label the point of intersection G.

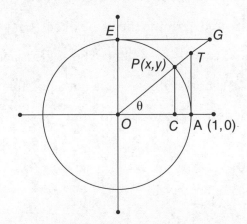

$\angle POC$ is complementary to $\angle EOG$ and $\angle EOG \cong \angle OPC$. With right angles $\angle OCP$ and $\angle OEG$, $\triangle OCP \sim \triangle GEO$. This gives the proportions $\dfrac{GO}{OE} = \dfrac{OP}{CP}$ and $\dfrac{OC}{PC} = \dfrac{EG}{OE}$. $\overline{EG}$ and $\overline{OG}$ are the tangent and secant of the triangle formed with the angle complementary to θ. Because of this, EG is the value of the cotangent function of theta, $\cot(\theta)$, and OG is the cosecant of theta, $\csc(\theta)$. Observe that $\dfrac{OC}{PC}$ is the reciprocal of $\dfrac{PC}{OC}$ and that $\dfrac{PC}{OC}$ is equal to $\tan(\theta)$. This means that $\cot(\theta)$ and $\tan(\theta)$ are reciprocals. Similarly, $\dfrac{OP}{CP}$ is the reciprocal of $\dfrac{CP}{OP}$ and $\dfrac{CP}{OP} = sin(\theta)$.

The relationships developed above are the beginning of equations called Trigonometric Identities. The **Reciprocal Identities** are:

$$\sec(\theta) = \frac{1}{\cos(\theta)}, \csc(\theta) = \frac{1}{\sin(\theta)}, \text{ and } \tan(\theta) = \frac{1}{\cot(\theta)}.$$

In addition to this, the triangles $\triangle OCP$, $\triangle OAT$, and $\triangle GEO$ are right triangles. Apply the Pythagorean Theorem to these triangles to $OC^2 + CP^2 = OP^2$, $OA^2 + AT^2 = OT^2$, and $OE^2 + EG^2 = OG^2$. Substituting the trigonometric values associated with the lengths of each of the given segments yields the **Pythagorean Identities**:

$$\sin^2(\theta) + \cos^2(\theta) = 1 \qquad 1 + \tan^2(\theta) = \sec^2(\theta) \qquad 1 + \cot^2(\theta) = \csc^2(\theta)$$

Note: the notation $\sin^2(\theta)$ means $(\sin(\theta))^2$.

We also have the **cofunctions identities**:

$$\sin(\theta) = \cos(90 - \theta) \qquad \sec(\theta) = \csc(90 - \theta) \qquad \tan(\theta) = \cot(90 - \theta)$$

> The nine identities listed in this section (with the double angle identities) are the most important identities for a student to know. (The double angle identities are not part of this course of study.)

EXAMPLE

If θ is an acute angle and $\sin(\theta) = \dfrac{3}{4}$ determine the values of the remaining 5 trigonometric functions.

Use the reciprocal identity to determine that $\csc(\theta) = \dfrac{4}{3}$. Use the Pythagorean Identity to determine the value of $\cos(\theta)$: $\left(\dfrac{3}{4}\right)^2 + \cos^2(\theta) = 1$ gives $\cos^2(\theta) = \dfrac{7}{16}$ and $\cos(\theta) = \dfrac{\sqrt{7}}{4}$. (Since the angle is acute, the x-coordinate of P is positive, so $\dfrac{-\sqrt{7}}{4}$ is not a possible answer.)

The other three functions are now easy to find:

$$\sec(\theta) = \frac{1}{\cos(\theta)} = \frac{4}{\sqrt{7}}$$

$$\tan(\theta) = \frac{\sin(\theta)}{\cos(\theta)} = \frac{3}{\sqrt{7}}$$

$$\cot(\theta) = \frac{1}{\tan(\theta)} = \frac{\sqrt{7}}{3}$$

EXAMPLE

> Given α (lowercase alpha) is an acute angle and tan(α) = 4. Find the values for the remaining 5 trigonometric functions.

> We know that $\cot(\alpha) = \dfrac{1}{4}$. Use the Pythagorean Identity for tangent: $1 + \tan^2(\alpha) = \sec^2(\alpha)$. $1 + 4^2 = \sec^2(\alpha)$ so $\sec(\alpha) = \sqrt{17}$ and $\cos(\alpha) = \dfrac{1}{\sqrt{17}}$. By definition, $\tan(\alpha) = \dfrac{\sin(\alpha)}{\cos(\alpha)}$ so $\tan(\alpha)\cos(\alpha) = \sin(\alpha)$ and $\sin(\alpha) = \dfrac{4}{\sqrt{17}}$. Finally, $\csc(\alpha) = \dfrac{\sqrt{17}}{4}$.

EXERCISE 9.1

In all cases for these exercises, the angle in question is an acute angle. Given the value of the indicated function for the angle, determine the value of the five other trigonometric angles for that angle.

1. $\cos(\alpha) = \dfrac{3}{5}$

2. $\csc(\beta) = \dfrac{13}{5}$

3. $\tan(\theta) = \dfrac{7}{24}$

4. $\sin(\omega) = \dfrac{40}{41}$

5. $\sec(\alpha) = \dfrac{3}{2}$

The Unit Circle—Beyond the First Quadrant

If we take the study of trigonometry onto the coordinate plane, we get to examine angles outside the interval [0, 90]. Moreover, we need to consider how an angle that rotates counterclockwise from point A differs from an angle that rotates clockwise. We designate any angle that is the result of a counterclockwise rotation as a positive angle and an angle formed by a clockwise rotation as negative.

Let's begin with positive angles first and leave the negative angles for later.

Second Quadrant. Take our angle θ on the unit circle and reflect it over the *y*-axis.

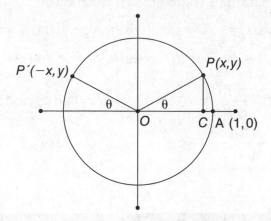

The measure of ∡*P'OA* = 180 − θ . The *y*-coordinate for both *P* and *P'* is *y*, so sin(180 − θ) = sin(θ). Notice cos(180 − θ) = − cos(θ). Can you explain why tan(180 − θ) = −tan(θ)? Since tan(θ) is the ratio of sin(θ) and cos(θ), tan(180 − θ) will be the ratio of sin(θ) and −cos(θ), which is equal to −tan(θ). Incidentally, the angle θ is called the **reference angle** for ∡*P'OA*. The reference angle is always the angle from the first quadrant that is associated with the pre-image of the transformation (*P* is the pre-image of *P'*).

Third Quadrant. Take our angle θ on the unit circle and reflect through the origin.

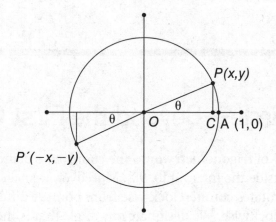

The measure of major angle ∡*P'OA* (the angle starting at *A* and drawn counterclockwise to *P*) is 180 + θ, and the reference angle is still θ. Both coordinates for *P'* are the negatives of the coordinates for *P*, so we get sin(180 + θ) = −sin(θ), cos(180 + θ) = −cos(θ), and tan(180 + θ) = tan(θ).

Fourth Quadrant. Take our angle θ on the unit circle and reflect across the x-axis.

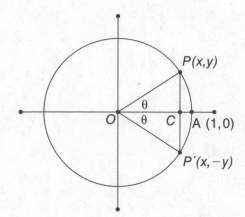

The measure of major angle $\angle P'OA = 360 - \theta$, and the reference angle is θ. The y-coordinates for P' is the negative of the y-coordinate for P, so we get $\sin(360 - \theta) = -\sin(\theta)$, $\cos(360 - \theta) = \cos(\theta)$, and $\tan(360 - \theta) = -\tan(\theta)$.

If the terminal side of the angle rotates clockwise into the fourth quadrant, you get the same result as if the terminal side was rotated into the fourth quadrant. Rather than refer to this angle as $360 - \theta$, the angle is designated as $-\theta$. Because these angles end in the same terminal angles, they are called **coterminal angles**. From the perspective of identities, we can see that:

$$\cos(-\theta) = \cos(\theta), \sin(-\theta) = -\sin(\theta), \text{ and } \tan(-\theta) = -\tan(\theta).$$

EXAMPLE

Determine the reference angles for each of the following:

(a) 160°

(b) 215°

(c) 290°

Solutions:

(a) 160° is in the second quadrant, so the reference angle is 180° – 160° = 20°

(b) 215° is in the third quadrant, so the reference angle is 215° – 180° = 35°

(c) 290° is in the fourth quadrant, so the reference angle is 360° – 290° = 70°.

(Remember: the reference angle is the acute angle formed by the terminal side of the angle and the x-axis.)

Express each of the following as a function of a positive acute angle:

(a) cos(160°)

(b) sin(215°)

(c) tan(290°)

Solutions:

(a) cos(160°) = −cos(20°)

(b) sin(215°) = −sin(35°)

(c) tan(290°) = −tan(70°)

Given $\cos(\theta) = a$, determine the value of $\tan(180 + \theta)$.

The terminal side of an angle with measure $180 + \theta$ is in the third quadrant. We know that both the sine and cosine functions are negative in this quadrant and that the reference angle is θ. Use the Pythagorean Identity to determine that $\sin(\theta) = \sin(\theta) = -\sqrt{1-a^2}$

so $\tan(\theta) = \dfrac{-\sqrt{1-a^2}}{-a} = \dfrac{\sqrt{1-a^2}}{a}$.

Even though technology has become used more and more, there are times when knowing the trigonometric values of these functions is important. Concentrate on learning the 30°, 45°, and 60° values and how the other angles use these as reference values.

There is one other issue that needs to be discussed, and that is the special triangles 45-45-90 and 30-60-90. A great deal of the work done in geometric-related problems still uses these angles and, as a consequence, their values have been included in the angles outside the first quadrant. In this day and age of technology being used, these values are considered old guard and should be part of one's base knowledge after studying trigonometry. The triangles:

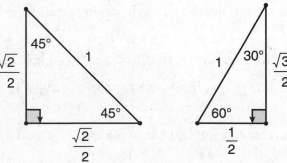

$$\sin\left(45^\circ\right) = \cos\left(45^\circ\right) = \frac{\sqrt{2}}{2} \qquad \sin\left(30^\circ\right) = \cos\left(60^\circ\right) = \frac{1}{2} \qquad \sin\left(60^\circ\right) = \cos\left(30^\circ\right) = \frac{\sqrt{3}}{2}$$

$$\tan\left(45^\circ\right) = 1 \qquad \tan\left(30^\circ\right) = \frac{\sqrt{3}}{3} \qquad \tan\left(60^\circ\right) = \sqrt{3}$$

EXERCISE 9.2

Questions 1–5: Express each of the following as a function of a positive acute angle.

1. sin(314°)

2. tan(−245°)

3. sec(268°)

4. csc(110°)

5. cos(−140°)

Question 6–10: Assume that θ is an acute angle.

6. Given cos(θ) = a, determine the value of sec(180 + θ).

7. Given sin(θ) = b, determine cot(360 − θ).

8. Given tan(θ) = c, determine the value of cos(180 − θ).

9. Given csc(θ) = d, determine the value of cos(−θ).

10. Given cot(θ) = e, determine the value of sin(−θ).

Complete the following chart with the exact values of the functions (no decimal approximations).

		sin(θ)	cos(θ)	tan(θ)	csc(θ)	sec(θ)	cot(θ)
11.	0°	0	1	−	1	−	0
12.	30°						
13.	45°						
14.	60°						
15.	90°						
16.	120°						
17.	135°						
18.	150°						
19.	180°						
20.	210°						
21.	225°						
22.	240°						
23.	270°						
24.	300°						
25.	315°						
26.	330°						
27.	360°						

Radian Measure

As the name hints at, the radian measure of an angle depends on the length of the radius of the circle in which it is formed. The ratio of the length of an arc of the circle to the length of the radius (each measure in the same units) gives the radian measure of the angle. In a circle with radius r, the circumference is given by $C = 2\pi r$. Therefore, an arc that forms a semicircle will have a radian value of $\dfrac{\pi r}{r} = \pi$ radians. That is, a 180° angle corresponds to π radians. We can convert any other angle, in either measurement, by using the ratio of 180° to π radians.

EXAMPLE

How many radians correspond to 65°?

Set up the proportion: $\dfrac{180}{\pi} = \dfrac{65}{\theta}$ and solve for θ: $180\theta = 65\pi$

so $\theta = \dfrac{65\pi}{180} = \dfrac{13\pi}{36}$.

EXAMPLE

Convert $\dfrac{\pi}{3}$ to degrees.

Set up the proportion: $\dfrac{180°}{\pi} = \dfrac{\theta}{\dfrac{\pi}{3}}$ and solve for θ: $\theta = \left(\dfrac{180°}{\pi}\right)\left(\dfrac{\pi}{3}\right) = 60°$.

An interesting aspect to this comes from our expectation that any measurement associated with the arc of a circle should contain a factor of π. The reason for this is that we often try to keep the degree measure of the angle as a "nice" number. Reread the definition of a radian: the angle needed so that the length of the arc of the circle is equal to the radius of the circle. So how many degrees correspond to 1 radian? The answer is $\theta = \dfrac{180}{\pi}$, and that is approximately 57.3°. You should know that the 30° angle corresponds to $\dfrac{\pi}{6}$ radians and a 45° angle corresponds to $\dfrac{\pi}{4}$ radians. You can find the other correspondences by multiplying these key relationships. The radian value equivalent to 135° is $\dfrac{3\pi}{4}$ because 135 = 3(45). Calculate $\sin\left(\dfrac{4\pi}{3}\right)$. Your initial reaction is to determine that $\dfrac{4\pi}{3}$ is $4\left(\dfrac{\pi}{3}\right) = 4(60) = 240$. You know that $\sin(240°)$ is $\dfrac{-\sqrt{3}}{2}$ so $\sin\left(\dfrac{4\pi}{3}\right) = \dfrac{-\sqrt{3}}{2}$.

The faster you learn where the angles are located on the unit circle, the easier it will be for you to evaluate trigonometric functions with the special angles. You know the number of radians in a full circle is 2π and half the circle is π. It stands to reason that a quarter circle is $\dfrac{\pi}{2}$ and three-fourths of the circle is $\dfrac{3\pi}{2}$. The angles 150° and 210° are each 30° from 180°. The 30° angle corresponds to $\dfrac{\pi}{6}$, so $150° = 180° - 30° = \pi - \dfrac{\pi}{6} = \dfrac{5\pi}{6}$ and 210° will correspond to $\dfrac{7\pi}{6}$ using a similar strategy.

EXERCISE 9.3

Questions 1–4: Convert each of the following into radian measure.

  1. 225°

  2. 720°

  3. –135°

  4. 144°

Questions 5–8: Convert each of the following into degree measure.

  5. $\dfrac{4\pi}{3}$

  6. $\dfrac{11\pi}{6}$

  7. $\dfrac{7\pi}{4}$

  8. $\dfrac{-2\pi}{3}$

Questions 9–15: Evaluate the trigonometric functions for the given value.

  9. $\sin\left(\dfrac{2\pi}{3}\right)$

  10. $\cos\left(\dfrac{3\pi}{4}\right)$

  11. $\tan\left(\dfrac{5\pi}{4}\right)$

  12. $\cot\left(\dfrac{4\pi}{3}\right)$

  13. $\sec\left(\dfrac{11\pi}{6}\right)$

  14. $\csc\left(\dfrac{-5\pi}{3}\right)$

  15. $\cos\left(\dfrac{3\pi}{4}\right)$

Graphs of Trigonometric Functions

Imagine point *P* beginning its path around the circle at point *A* and rotating counterclockwise around the circle at a rate of 1 unit per second. The needed to travel the distance around the circle will take 2π seconds at which point the values will begin to repeat themselves.

We'll use the diagram of the unit circle with the segments whose lengths form the six trigonometric functions to help us understand the graphs of the functions.

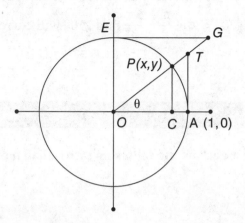

The function $s(\theta) = \sin(\theta)$ has an initial value of 0. After $\dfrac{\pi}{2}$ seconds, *P* is at the top of the circle and the value of $s(\theta) = 1$.

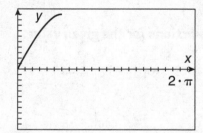

On the interval $\left[\dfrac{\pi}{2}, \pi\right]$ seconds, *P* works its way back to the *x*-axis.

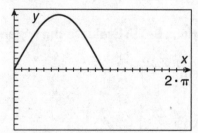

On the interval $\left[\pi, \dfrac{3\pi}{2}\right]$ seconds, *P* works its way to the bottom of the circle and its minimum value of –1.

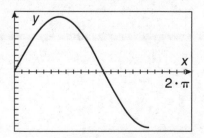

The last portion of the trip takes us back to the *x*-axis during the interval $\left[\dfrac{3\pi}{2}, 2\pi\right]$ seconds.

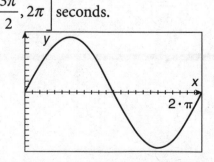

Should we allow P to keep traveling, the graph would begin to repeat itself. The domain of the function $f(x) = \sin(x)$ is the set of real numbers, $(-\infty, \infty)$, and the range is $[-1, 1]$.

The function $c(\theta) = \cos(\theta)$ has an initial value of 0. After $\dfrac{\pi}{2}$ seconds, P is at the top of the circle and the value of $c(\theta) = 1$.

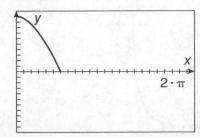

On the interval $\left[\dfrac{\pi}{2}, \pi\right]$ seconds the graph continues to decrease until it reaches the minimum value of $c(\theta)$ at -1.

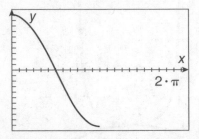

The graph begins to increase on the interval $\left[\pi, \dfrac{3\pi}{2}\right]$ seconds as P returns to the y-axis.

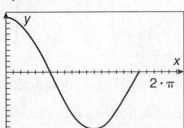

Finally, P returns to the x-axis during the interval $\left[\dfrac{3\pi}{2}, 2\pi\right]$ seconds.

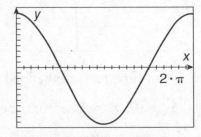

The domain of the function $f(x) = \cos(x)$ is $(-\infty, \infty)$ and the range is $[-1, 1]$. In both cases, the essential shapes of the sine and cosine graphs can be seen by plotting the five points associated with $\theta = 0, \dfrac{\pi}{2}, \pi, \dfrac{3\pi}{2}$, and 2π. We'll see more of this in the next section.

The graphs of the cosecant and secant functions are more easily determined by the reciprocal nature to the sine and cosine functions than they are to the graph of the unit circle and the associated segments. There are just two things to keep in mind: the reciprocal of a small number is a big number and the reciprocal of 0 is undefined. As we saw with the rational functions, when a function contains a point in an interval for which the function is undefined, the graph will contain an asymptote. The cosecant function is undefined whenever θ is equal to a multiple of π. As the $\sin(\theta)$ gets closer to 0, the $\csc(\theta)$ will get infinitely large and will have the same sign as $\sin(\theta)$.

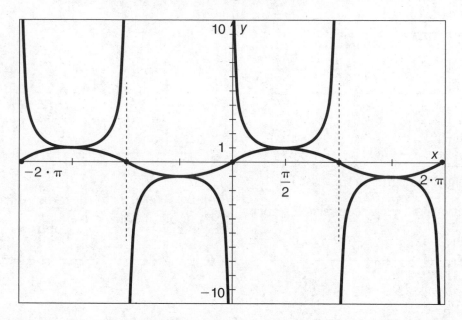

The secant function is undefined whenever θ is equal to an odd multiple of $\frac{\pi}{2}$. As the $\cos(\theta)$ gets closer to 0, the $\sec(\theta)$ will get infinitely large and will have the same sign as $\cos(\theta)$.

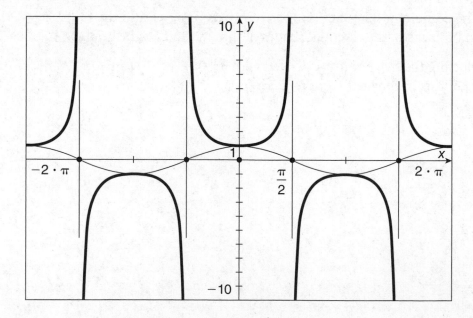

The graphs of the tangent and cotangent functions will also contain asymptotes, with the graph of the tangent function having the same asymptotes as the graph of the secant function and the graph of the cotangent function having the same asymptotes as the graph of the cosecant function.

The graph of $y = \tan(\theta)$

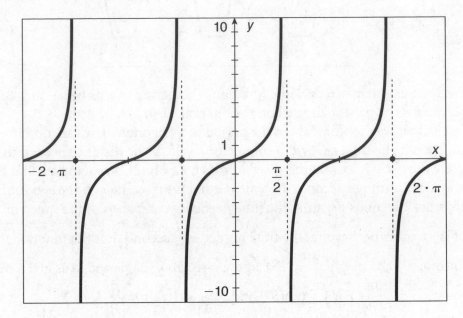

The graph of $y = \cot(\theta)$

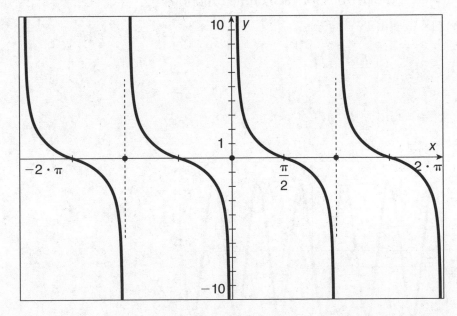

You've noticed, no doubt, that these graphs repeat themselves after a certain interval. The size of the interval is called **the period** of the function. The period for the sine, cosine, cosecant, and secant functions is 2π, while the period is π for the tangent and cotangent functions. In general, the period, p, of a function is the value for which $f(x) = f(x + p)$ for all values of x.

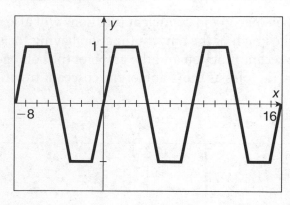

This is an example of a periodic function. The period of this function is 8. (Observe how the pattern formed on the interval [0, 8] is repeated.)

An efficient way to think about the period of a trigonometric function is to think about the time needed to complete a cycle. If the distance needed to complete the cycle is 2π and the speed at which an object is moving around the unit circle is 1 unit per second, the time needed is 2π seconds. If the speed of the object is 2 seconds per unit, the time needed is π seconds. If the speed of the object is 4π units per second, the time needed is $\frac{1}{2}$ second. For the functions $y = \sin(Bx)$, $y = \cos(Bx)$, $y = \csc(Bx)$, and $y = \sec(Bx)$, the period, p, is given by the equation $p = \frac{2\pi}{|B|}$. For $y = \tan(Bx)$ and $y = \cot(Bx)$, $p = \frac{\pi}{|B|}$.

EXAMPLE

Determine the periods for each of the functions:

(a) $f(x) = \sin(4x)$

(b) $g(x) = \cos\left(\dfrac{\pi x}{3}\right)$

(c) $k(x) = \tan\left(\dfrac{x}{3}\right)$

(d)

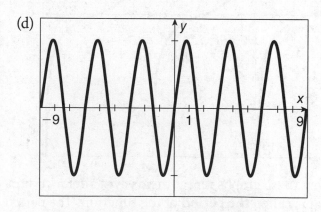

Solutions:

(a) $p = \dfrac{2\pi}{4} = \dfrac{\pi}{2}$

(b) $p = \dfrac{2\pi}{\dfrac{\pi}{3}} = 6$

(c) $p = \dfrac{\pi}{\dfrac{1}{3}} = 3\pi$

(d)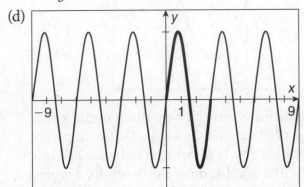

As you can see in the graph, a complete cycle is completed in 3 units.
The period is equal to 3.

Write the equation for the graph shown in part (d) of the previous example.

The period is equal to 3, so $3 = \dfrac{2\pi}{|B|}$. Solve for B: $|B| = \dfrac{2\pi}{3}$. The maximum value of the sine function appears a quarter of the way through the period so $B > 0$. Therefore, the equation for the function is $y = \sin\left(\dfrac{2\pi}{3}x\right)$.

How does the graph of $y = -2\sin(x)$ differ from the graph of $y = \sin(x)$?

The graph of $y = 2f(x)$ differs from the graph of $y = f(x)$ in that the graph is stretched from the x-axis by a factor of 2. The graph of $y = -f(x)$ differs from the graph of $y = f(x)$ in that the graph is reflected over the x-axis.

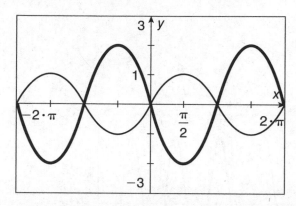

> **The five key points for graphing a sine or cosine function are the beginning of the interval, one-fourth of the way through the interval, halfway through the interval, three-fourths of the way through the interval, and the end of the interval forming one period of the graph.**

The graphs of all sine and cosine functions can be written in the form $f(x) = A\sin(Bx + C) + D$ and $f(x) = A\cos(Bx + C) + D$. The impact of C is to translate horizontally, but this is something to be studied in a future course in trigonometry. The **amplitude** of the function, designated by $|A|$, is half the difference of the maximum and minimum values of the function. As we saw earlier, $|B|$ is used to determine the period of the function. The parameter D describes the vertical translation of the graph. (The value D is also called the **average** value of the function.) Because the other four trigonometric functions have ranges that go to infinity, we do not claim that they have an amplitude.

The amplitude of the graph of $y = -2\sin(x)$ is 2. The maximum value is 2 and the minimum value is -2, so $A = \dfrac{2-(-2)}{2} = 2$.

Determine the amplitude of the graph of a cosine function whose maximum value is 9 and whose minimum value is -3.

Amplitude $= \dfrac{9-(-3)}{2} = 6$.

Sketch, without the use of a graphing utility, the graph of
$y = 4\cos\left(\dfrac{2}{3}x\right) + 2$ over one period.

▶ The period of the function is $\dfrac{2\pi}{\frac{2}{3}} = 3\pi$. The points of interest are when

$x = 0, \dfrac{3\pi}{4}, \dfrac{3\pi}{2}, \dfrac{9\pi}{4}$, and 3π. These represent the initial value, one-fourth of the way through the cycle, halfway through the cycle, three-fourths of the way through the cycle, and the end of the cycle. The coordinates of these points are $(0, 6)$, $\left(\dfrac{3\pi}{4}, 2\right)$, $\left(\dfrac{3\pi}{2}, -2\right)$, $\left(\dfrac{9\pi}{4}, 2\right)$, and $(3\pi, 6)$. (Observe that the y-coordinates are, in order, the maximum, average, minimum, average, and maximum values of the function.) Having the coordinates, plot these points.

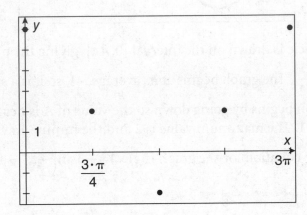

▶ Draw the cosine graph through these points.

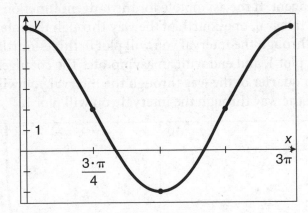

EXAMPLE

Write an equation for the function $g(x)$ whose graph over one period is shown.

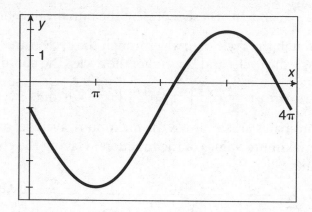

A complete cycle is drawn on the interval $[0, 4\pi]$, giving the period as 4π and $B = \dfrac{2\pi}{4\pi} = \dfrac{1}{2}$. The graph begins at its average, –1, so it is a sine function and the motion begins by going down so the value of A is negative and the value of D is –1. The maximum value is 2 and the minimum value is –4, so $|A| = 3$. The equation for the graph is $g(x) = -3\sin\left(\dfrac{1}{2}x\right) - 1$.

It is a little more challenging to describe how to sketch the graph of the tangent and cotangent. If the asymptote for the tangent function is at the beginning of the interval, one-fourth of the way through the interval you will plot –1, halfway through the interval you will plot 0, three-fourths of the way through you will plot 1, and end with an asymptote. The cotangent graph is different in that a quarter of the way through the interval you will plot 1 and three-quarters of the way through the interval you will plot 1.

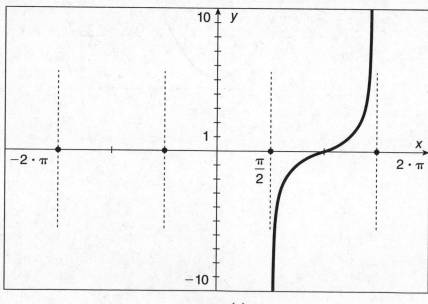

$$y = \tan(x)$$

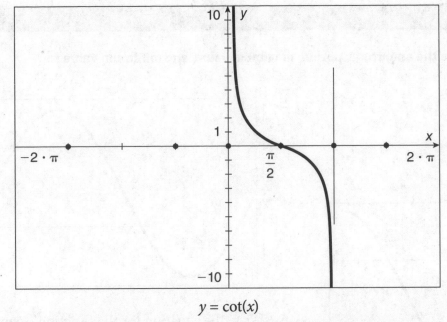

$$y = \cot(x)$$

When drawing a secant or cosecant graph, first replace the trigonometric function with its reciprocal, draw that function, include asymptotes whenever this adjusted function is zero, and then sketch the original function (working from the maximum and minimum values and going toward the asymptotes).

EXAMPLE

▶ Sketch the graph of $y = 4\sec\left(\dfrac{2}{3}x\right) + 2$ over one period.

▶ Replace sec with cos. Sketch $y = 4\cos\left(\dfrac{2}{3}x\right) + 2$. We did that just a moment ago. Sketch in the asymptotes. Sketch in the function.

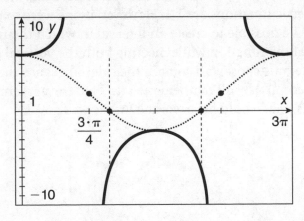

EXERCISE 9.4

For questions 1–5: Determine the amplitude, period, maximum value, and minimum value for each function.

1. $f(x) = 8\sin(4x) + 5$

2. $g(x) = 5\cos(4\pi x) + 3$

3. $k(x) = -4\sin\left(\dfrac{\pi}{2}x\right) - 1$

4.

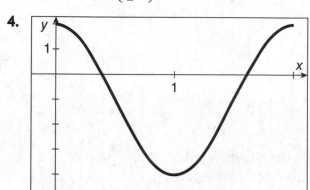

5.

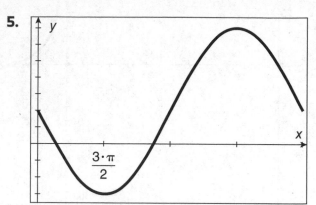

6. Write the equation for the function drawn in question 4.

7. Write the equation for the function drawn in question 5.

Inverse Trigonometric Functions

A function must be 1-1 in order for the function to have an inverse. None of the six trigonometric functions are 1-1. However, if one were to restrict the domain, it would be possible to create an interval in which the function covers the entire range of the function while meeting both the vertical and horizontal line tests. The intervals chosen all contain the origin because they can. This is a convention accepted by all mathematicians. Had the agreement been for different intervals, that would have worked just as well.

Function	Domain	Range	Graph
$y = \sin(x)$	$\left[\dfrac{-\pi}{2}, \dfrac{\pi}{2}\right]$	$[-1, 1]$	
$y = \sin^{-1}(x)$	$[-1, 1]$	$\left[\dfrac{-\pi}{2}, \dfrac{\pi}{2}\right]$	
$y = \cos(x)$	$[0, \pi]$	$[-1, 1]$	
$y = \cos^{-1}(x)$	$[-1, 1]$	$[0, \pi]$	
$y = \tan(x)$	$\left(-\dfrac{\pi}{2}, \dfrac{\pi}{2}\right)$	$(-\infty, \infty)$	
$y = \tan^{-1}(x)$	$(-\infty, \infty)$	$\left(-\dfrac{\pi}{2}, \dfrac{\pi}{2}\right)$	

Example: Evaluate $\sin^{-1}\left(\dfrac{\sqrt{3}}{2}\right)$.

Solution: The notation asks, "What is the angle whose sine is $\dfrac{\sqrt{3}}{2}$?" The answer is $\dfrac{\pi}{3}$.

Because the definition of a radian angle relates the angle size to the length of the corresponding arc in the unit circle, the notation can also be thought of as asking, "What is the arc whose length ends at a point whose y-coordinate is $\dfrac{\sqrt{3}}{2}$?" the notation arcsin is used rather than $\sin^{-1}$.

Example: Evaluate $\sin^{-1}\left(\dfrac{-\sqrt{3}}{2}\right)$.

Solution: This question gets to the heart of the issue. There are two places on the unit circle where the y-coordinate is $\dfrac{-\sqrt{3}}{2}$, and there are an infinite number of ways to get to that point. The range of the inverse sine *function* is $\left[\dfrac{-\pi}{2}, \dfrac{\pi}{2}\right]$, and there is only one value in this interval for which the sine function is $\dfrac{-\sqrt{3}}{2}$ and that value is $\dfrac{-\pi}{3}$.

An important distinction: What is the difference between solving the equation $x^2 = 4$ and evaluating $\sqrt{4}$? When solving an equation, you are expected to give all values for which the equation is true. When evaluating a function, you are expected to give the *one* value for which the expression is true. The solution to the equation $\sin(x) = \dfrac{-\sqrt{3}}{2}$ on the interval $[0, 2\pi]$ is $\dfrac{4\pi}{3}$ and $\dfrac{5\pi}{3}$, whereas $\sin^{-1}\left(\dfrac{-\sqrt{3}}{2}\right) = \dfrac{-\pi}{3}$.

Example: Evaluate $\cos^{-1}\left(\dfrac{-\sqrt{2}}{2}\right)$.

Solution: The range of the inverse cosine function is $[0, \pi]$ so $\cos^{-1}\left(\dfrac{-\sqrt{2}}{2}\right) = \dfrac{3\pi}{4}$.

EXAMPLE

Example: Evaluate $\tan^{-1}\left(\dfrac{-\sqrt{3}}{3}\right)$.

Solution: The range of the inverse tangent function is $\left(-\dfrac{\pi}{2}, \dfrac{\pi}{2}\right)$ so $\tan^{-1}\left(\dfrac{-\sqrt{3}}{3}\right) = \dfrac{-\pi}{6}$.

EXAMPLE

Example: Evaluate $\sin\left(\tan^{-1}\left(\dfrac{-\sqrt{3}}{3}\right)\right)$.

Solution: You just determined that $\tan^{-1}\left(\dfrac{-\sqrt{3}}{3}\right) = \dfrac{-\pi}{6}$ so the problem becomes an evaluation of $\sin\left(\dfrac{-\pi}{6}\right)$ and this, you know, is $\dfrac{-1}{2}$.

EXAMPLE

Example: Evaluate $\cos\left(\sin^{-1}\left(\dfrac{5}{13}\right)\right)$.

Solution: *This* is different. $\dfrac{5}{13}$ is not a point we saw when we looked at the unit circle. When problems seem difficult and the means to the solution are not clear, it is often best to go back to basics. $\sin^{-1}\left(\dfrac{5}{13}\right)$ is an angle. Let's call that angle θ. Once we find θ, we need to determine $\cos(\theta)$. We have $\sin(\theta) = \dfrac{5}{13}$ and want to determine $\cos(\theta)$. We can use the Pythagorean Identity $\sin^2(\theta) + \cos^2(\theta) = 1$ or we can draw a right triangle with hypotenuse 13 and leg opposite θ with length 5. You know the third side of the triangle has length 12 so the answer to $\cos\left(\sin^{-1}\left(\dfrac{5}{13}\right)\right)$ is $\dfrac{12}{13}$.

Example: Evaluate $\tan\left(\sin^{-1}\left(\dfrac{-24}{25}\right)\right)$.

Solution: Let $\theta = \sin^{-1}\left(\dfrac{-24}{25}\right)$ so that $\sin(\theta) = \dfrac{-24}{25}$. We could use the Pythagorean Identity $\sin^2(\theta) + \cos^2(\theta) = 1$ to find $\cos(\theta)$ and then determine the value of $\tan(\theta)$ by using the ratio of $\sin(\theta)$ and $\cos(\theta)$ or we can use the $\csc(\theta)$ is the reciprocal of $\sin(\theta)$ followed by the Pythagorean Identity $1 + \cot^2(\theta) = \csc^2(\theta)$ to find the value of $\cot(\theta)$ and then take the reciprocal of this response to get the value of $\tan(\theta)$. (Exhale—that was a lot to think about.) It would probably be just as easy to use the Pythagorean Theorem to determine the third side of the triangle is 7. *But*—what do we know about θ? The inverse sine is a negative value, so the angle in question must be between $\dfrac{-\pi}{2}$ and 0. In this interval, $\tan(\theta)$ is negative, so $\tan\left(\sin^{-1}\left(\dfrac{-24}{25}\right)\right) = \dfrac{-24}{7}$. Drawing the picture would be very helpful here.

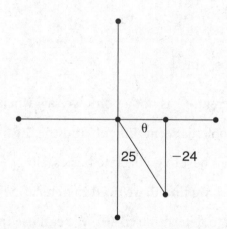

Labeling the leg of this triangle as –24 emphasizes the quadrant in which the terminal side of the angle is located.

EXERCISE 9.5

Evaluate each of the following.

1. $\tan^{-1}\left(-\sqrt{3}\right)$

2. $\sin^{-1}\left(\dfrac{-\sqrt{2}}{2}\right)$

3. $\sec^{-1}(-2)$

4. $\sin\left(\cos^{-1}\left(\dfrac{-8}{17}\right)\right)$

5. $\tan\left(\cos^{-1}\left(\dfrac{-3}{4}\right)\right)$

Solving Trigonometric Equations

It is traditional at this level of study to ask you to solve trigonometric equations limiting the solutions to the interval [0, 360°] or [0, 2π]. However, be sure you read the directions of the problem to ensure you give the answers being sought.

SOLUTIONS IN DEGREE MEASURE

Be sure to set your calculator to degree mode.

EXAMPLE

Solve: $3\tan(\theta) - 4 = 0$ over the interval [0, 360°]. Answer to the nearest tenth of a degree.

Add 4 and divide by 3 to determine $\tan(\theta) = \dfrac{4}{3}$. Use the inverse tangent command to determine $\theta = 53.1°$. We have to remember that the inverse tangent function gives answers between –90° and 90° (the degree equivalent to $\left(\dfrac{-\pi}{2}, \dfrac{\pi}{2}\right)$ radians). The tangent function is positive in the third quadrant. Using 53.1° as the reference angle, the third quadrant angle for which $\tan(\theta) = \dfrac{4}{3}$ is 180° + 53.1° = 233.1°. The answer to the problem is θ = 53.1°, 233.1°.

EXAMPLE

Solve: $6\sin^2(\beta) + 5\sin(\beta) - 1 = 0$ over the interval [0, 360°]. Answer to the nearest tenth of a degree.

Factor the quadratic: $(6\sin(\beta) - 1)(\sin(\beta) + 1) = 0$. Set each factor equal to 0 and solve: $\sin(\beta) = \dfrac{1}{6}, -1$. $\sin(\beta) = -1$ when β = 270°. Use the inverse sine command to solve $\sin(\beta) = \dfrac{1}{6}$ to get β = 9.6°. Remembering that the sine function is positive in the second quadrant as well as the first quadrant, the other solution to this problem is 180° – 9.6° = 170.4°. The solution to this problem is β = 9.6°, 170.4°, and 270°.

▶ Solve: $3\tan^2(\beta) + 5\tan(\beta) - 10 = 0$ over the interval [0, 360°]. Answer to the nearest tenth of a degree.

▶ Do we try to factor the quadratic or not? The discriminant for this quadratic is $5^2 - 4(3)(-10) = 145$. This is not a perfect square, so the roots to the equation will be irrational. Use the quadratic formula to solve this problem, keeping in mind that the variable of the quadratic is $\tan(\beta)$,

$$\tan(\beta) = \frac{-5 \pm \sqrt{145}}{6} = 1.1736, -2.84027.$$ (Use the memory feature of your calculator to store the decimal approximation of the solution.)

▶ The first solution to this problem is $\tan^{-1}(A) = 49.6°$. The third quadrant angle with this reference angle is 229.6°.

▶ The solutions for the angle whose tangent values are negative are in the second and fourth quadrants. We find the reference for these angles by using 2.840265763. To do this, we'll enter $\tan^{-1}(-B)$ in the calculator and store this value in memory location C.

NORMAL FLOAT AUTO REAL DEGREE MP

$(-5+\sqrt{145})/6 \rightarrow A$

                          1.173599096

$(-5-\sqrt{145})/6 \rightarrow B$

                          -2.840265763

■

NORMAL FLOAT AUTO REAL DEGREE MP

$\tan^{-1}(1.173599096)$

                          49.56635603

$\tan^{-1}(--2.840265763) \rightarrow C$

                          70.60386707

▶ The reference angle is 70.6°. The second quadrant angle is 109.4° and the fourth quadrant angle is 289.4°. The solution to the equation $3\tan^2(\beta) + 5\tan(\beta) - 10 = 0$ is $\beta = 49.6°, 109.4°, 229.6°,$ and 289.4°.

SOLUTIONS IN RADIAN MEASURE

Be sure to set your calculator to radian mode.

EXAMPLE

▶ Solve $\sin^2(\theta) = \dfrac{1}{4}$ with $\theta \in [0, 2\pi]$.

▶ Solution: Take the square root of both sides of the equation to get $\sin(\theta) = \pm\dfrac{1}{2}$. $\sin(\theta) = \dfrac{1}{2}$ when $\theta = \dfrac{\pi}{6}, \dfrac{5\pi}{6}$ and $\sin(\theta) = \dfrac{-1}{2}$ when $\theta = \dfrac{7\pi}{6}, \dfrac{11\pi}{6}$. Therefore, the solution is $\theta = \dfrac{\pi}{6}, \dfrac{5\pi}{6}, \dfrac{7\pi}{6}, \dfrac{11\pi}{6}$.

EXAMPLE

▶ Solve $2\cos^3(\theta) - \cos(\theta) = 0$ with $\theta \in [0, 2\pi]$.

▶ Factor $\cos(\theta)$ to get $\cos(\theta)(2\cos^2(\theta) - 1) = 0$. Set each factor equal to 0, $\cos(\theta) = 0$ or $2\cos^2(\theta) - 1 = 0$.

▶ $\cos(\theta) = 0$ gives $= \dfrac{\pi}{2}, \dfrac{3\pi}{2}$

▶ $2\cos^2(\theta) - 1 = 0$ becomes $\cos^2(\theta) = \dfrac{1}{2}$ so $\cos(\theta) = \pm\sqrt{\dfrac{1}{2}} = \pm\dfrac{\sqrt{2}}{2}$. Consequently, $\theta = \dfrac{\pi}{4}, \dfrac{3\pi}{4}, \dfrac{5\pi}{4}, \dfrac{7\pi}{4}$.

▶ The solution to the problem is $\theta = \dfrac{\pi}{4}, \dfrac{3\pi}{4}, \dfrac{5\pi}{4}, \dfrac{7\pi}{4}, \dfrac{\pi}{2}, \dfrac{3\pi}{2}$.

EXAMPLE

▶ Solve $\sec^2(\theta) - 3\sec(\theta) + 2 = 0$ over the interval $[0, 2\pi]$.

▶ The format of this equation is just like $x^2 - 3x + 2 = 0$. Factor the trinomial to $(\sec(\theta) - 2)(\sec(\theta) - 1) = 0$. Set each factor equal to 0: $(\sec(\theta) - 2) = 0$ or $(\sec(\theta) - 1) = 0$ and solve for $\sec(\theta)$: $\sec(\theta) = 2$ or $\sec(\theta) = 1$. Most people do not have the values of the secant function memorized, but they do know that $\sec(\theta)$ and $\cos(\theta)$ are reciprocals. $\sec(\theta) = 2$ or $\sec(\theta) = 1$ becomes $\cos(\theta) = \dfrac{1}{2}$ or $\cos(\theta) = 1$ from which it can be determined that $\theta = 0, \dfrac{\pi}{3}, \dfrac{5\pi}{3}, 2\pi$.

Solve $2\sin^2(\theta) + 3\cos(\theta) = 0$ over the interval $[0, 2\pi]$.

We are going to use the Pythagorean Identity $\sin^2(\theta) + \cos^2(\theta) = 1$ and rewrite $\sin^2(\theta) = 1 - \cos^2(\theta)$. Substitute $1 - \cos^2(\theta)$ for $\sin^2(\theta)$ in the original equation to get the equivalent equation:

$$2\left(1 - \cos^2(\theta)\right) + 3\cos(\theta) = 0$$

Distribute through the parentheses:

$$2 - 2\cos^2(\theta) + 3\cos(\theta) = 0$$

Simplify:

$$-2\cos^2(\theta) + 3\cos(\theta) + 2 = 0$$

Multiply both sides of the equation by –1:

$$2\cos^2(\theta) - 3\cos(\theta) - 2 = 0$$

Factor:

$$\left(2\cos(\theta) + 1\right)\left(\cos(\theta) - 2\right) = 0$$

Set each factor equal to 0 and solve:

$$\cos(\theta) = \frac{-1}{2}, 2$$

There is no solution to $\cos(\theta) = 2$ because 2 is outside the range of the cosine function. Therefore, $\theta = \dfrac{2\pi}{3}, \dfrac{4\pi}{3}$ is the solution to the problem.

> It isn't often done at this level, but you could be asked to solve an equation in quadratic form with the angles measured in radian value rather than degree.

Solve: $3\tan^2(\beta) + 5\tan(\beta) - 10 = 0$ over the interval $[0, 2\pi]$. Answer to the nearest hundredth of a radian.

We have that $\tan(\beta) = \dfrac{-5 \pm \sqrt{145}}{6}$ from using the quadratic formula.

Change the mode of your calculator from degrees to radians. Repeat the keystrokes to get the reference angles.

```
NORMAL FLOAT AUTO REAL RADIAN MP          🔋

tan⁻¹ (A)
                                   0.8650961112
- - - - - - - - - - - - - - - - - - - - - - - - -
tan⁻¹ (−B)
                                    1.232269945
- - - - - - - - - - - - - - - - - - - - - - - - -
Ans → C
                                    1.232269945
- - - - - - - - - - - - - - - - - - - - - - - - -
■
```

▶ The third quadrant angle is 0.865 plus π: 4.01. The second quadrant angle is π − the reference angle: π − 1.232 = 1.91, and the fourth quadrant angle is 2π − the reference angle = 5.05. The solution to $3\tan^2(\beta)+5\tan(\beta)-10=0$ is β = 0.87, 1.91, 4.01, and 5.05.

EXERCISE 9.6

Solve these equations on the interval [0, 360°]. Give answers to the nearest tenth of a degree.

1. $5\cos(\alpha)+3=0$

2. $\tan^2(\beta)-\tan(\beta)-2=0$

3. $9\sin^2(\omega)-4=0$

4. $12\cos^2(A)-\cos(A)-6=0$

5. $3\sec^2(Q)-17\sec(Q)-5=0$

Solve these equations on the interval $\left[0,2\pi\right]$. Give answers to the nearest hundredth of a radian.

6. $4\sin(\alpha)+3=0$

7. $9\tan^2(\beta)-4=0$

8. $5\sin^2(R)-2\sin(R)-3=0$

9. $\csc^2(\beta)-5\csc(\beta)+4=0$

10. $6\tan^2(\theta)-5\tan(\theta)-4=0$

Applications of Periodic Functions

The height of a seat on a Ferris wheel above the ground as the wheel rotates about its axis, the depth of tidal water at a given hour, and regional temperature throughout the year are examples of periodic functions. The graphs that display these phenomena are called **sinusoidal waves** because the shape is either a sine or cosine.

▶ The diameter of a Ferris wheel is 80 feet. The maximum height of a seat on this wheel is 100 feet above the ground. Given that the wheel completes 5 revolutions in 6 minutes, write an equation for the height above ground, in feet, of a seat that begins the ride at its maximum height t seconds after the ride begins.

▶ Let's sort out what we know. The seat begins at its maximum position, decreases to its lowest position, and then returns to its highest position. This describes the graph of the cosine. (Note that the ride does not begin until all the seats have been filled, so we do not count the time it takes to occupy the seat, then rotate slightly so the next seat can be filled repeatedly until all the seats are filled.)

▶ The maximum height of the wheel is 100 feet, and the diameter of the wheel is 80 feet. This makes the minimum height of the wheel 20 feet and the center of the wheel 60 feet above the ground. The amplitude of the graph is $\dfrac{100-20}{2} = 40 = 40$ feet, the radius of the wheel. We have the equation for the motion of the seat as $h(t) = 40\cos(Bt) + 60$. The wheel makes 5 revolutions in 6 minutes (360 seconds) so the wheel makes one revolution, the period of the graph, 72 seconds. Solve for B: $72 = \dfrac{2\pi}{B}$ so $B = \dfrac{\pi}{36}$. The equation representing the height of the seat is $h(t) = 40\cos\left(\dfrac{\pi}{36}t\right) + 60$.

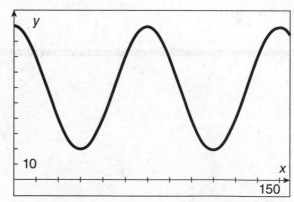

EXAMPLE

Regional temperatures are predicted based on data that has been collected over a number of decades. The typical highest temperature in Charlotte, North Carolina, is 86° and occurs on July 18, while the typical lowest temperature, 46°, occurs on January 18. If we assume that these temperatures occur in a periodic manner, write the equation illustrating the typical temperature in Charlotte, North Carolina. (These are typical temperatures, not any particular year. Also, as they are based on decades of data, any climate changes that have occurred in the last decade will not appear to change these values too drastically. The impact of climate change will not impact this function for another decade.)

We have a period of 365 days, a maximum value of 86°, and a minimum of 46°. We will begin our function on the eighteenth of January (calling this day zero). The amplitude of the function is $\dfrac{86° - 46°}{2} = 20°$ and the average value is 66°. Find the last parameter for the equation: $B = \dfrac{2\pi}{365}$ so the equation for the typical temperature, F, in Charlotte, North Carolina, is given by $F(d) = -20\cos\left(\dfrac{2\pi}{365}d\right) + 66$, where d is the number of days after January 18.

Do you see why the leading coefficient is negative? The graph of the cosine usually begins with its maximum value. This representation begins with its minimum (or a reflection across the x-axis) and thus negates the leading coefficient.

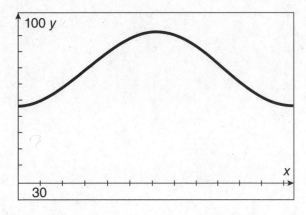

EXERCISE 9.7

Follow the instructions for each of the problems below.

1. The High Roller Ferris wheel in Las Vegas, Nevada is the world's tallest Ferris wheel. The diameter of the wheel is 520 feet and the wheel reaches a maximum height of 550 feet during its 30-minute rotation. Passengers get on a car at the bottom of its rotation as the car moves at 1 foot per second. Write an equation for the height, h, of a passenger t seconds after entering a car.

2. The depth of the water in the Bay of Fundy follows a sinusoidal pattern with the high tide being 53 feet and low tide being 11 feet. The time between low tide and high tide is 12 hours 25 minutes. Write an equation for the depth, d, of the water t minutes after low tide is reached.

3. Los Angeles, California, typically reaches its highest temperature, 84°, around August 20, and its lowest temperature, 68°, about January 4. Assuming the temperature for this region follows a sinusoidal pattern, write an equation for the typical temperature, T, of Los Angeles d days after January 4.

4. An exercise used by submarine captains is something called "porpoising." When in danger, the captain will order the submarine to oscillate in a sinusoidal manner between shallow and extreme depths. Suppose the submarine is at a depth of 130 m when the captain gives the order to begin porpoising, starting with moving to depth of 10 m and then to a depth of 250 m. Write an equation to determine the depth, d, of the submarine t minutes after the order is given if the time needed to complete one cycle is 5 minutes.

CHAPTER 10

Descriptive Statistics

There are two branches of statistics—descriptive and inferential. You have seen both of these in use. If you watch a sporting event, players and team statistics are plentiful—batting averages, yards per pass attempt, percent of first serves in play, free throw percentages, average saves per game. Examples from outside the sports world include number of ounces of fruit juice in a bottle, miles per gallon used during a long trip, and the number of hours spent each night doing homework.

All of these examples describe some quantity. Inferential statistics are used to gauge something about a population based on data gathered from a subset (called a sample) of the whole. Who will win the election? Surveys indicate that Candidate M has 54% of the vote with a maximum error of 3%. A manufacturer claims its soft drink dispenser "pours 12 ounces into a cup" each time it is used. A restaurant owner who uses such a dispenser in her store says this is not so and that the average number of ounces delivered is less than 12 ounces. We will examine some of the aspects of descriptive statistics in this chapter and will examine the basics of inferential statistics in the next chapter.

Measures of Central Tendency

The average of a set of data represents a value associated with the center of the data. There are three measures of central tendency: mode, median, and mean. The **mode** of a set of data is the value (values) that occurs with the highest frequency. It is possible for a set of data to have more than one mode. As a rule, the mode is the least used measure of center.

Mean, mode, and median can all be referred to as the average.

The **median** of a set of data is used for data that has been sorted from least to largest or vice versa. If the number of pieces of data is odd, the median is that piece equally distant from either end. For the sorted data 14, 21, 23, 29, 37, 53, 100 the median is 29 as it is fourth in place from the lowest value or the highest. If there is an even number of pieces of data, the median is one-half the sum of the two numbers in the middle. For the sorted data 14, 14, 21, 23, 29, 37, 53, 100 the median is $\frac{29+37}{2} = 33$. It is worth noting that the median does not have to be one of the data values. (Did you notice that the first data set had no mode because all values appear the same number of times and that the mode of the second data set is 14 because it occurs twice and all other values occur only once?)

The **mean**, the measure that is most associated with the notion of average, is equal to the sum of the data values divided by the number of pieces of data, that is, mean $= \dfrac{\sum\limits_{i=1}^{n} x_i}{n}$. The uppercase Greek letter sigma, Σ, indicates a summation process, while the subscripted variable x_i is used to represent each of the data values.

> **EXAMPLE**
>
> Find the mean of the data: 14, 14, 21, 23, 29, 37, 53, 100
>
> Mean $= \dfrac{291}{8} = 36.375$

When looking at sets of data, you'll need to determine if the data represents all the possible data values (called the **population**) or a subset of all the data values (called a **sample**). The conclusions drawn for a population are called **parameters**, and they are represented by lowercase Greek letters. The symbol for the parameter mean is mu, μ. The conclusions drawn for a sample are **statistics**. The symbol for the statistic mean is called x-bar, $\bar{x}$. Assuming the data given in this example is a sample, we have $\bar{x} = 36.375$. The mean and median for this data set are fairly close to each other.

> **EXAMPLE**
>
> Take the data set from the last example and include one more piece of data, 500. Compute the median and the mean.
>
> 14, 14, 21, 23, 29, 37, 53, 100, 500
>
> There are 9 pieces of data in the set and the middle (fifth) piece is the median, 29. The sum of the nine pieces of data is 791. The mean of the data is $\dfrac{791}{9} = 87.9$.

There is a large gap between these two answers. The median is not impacted by how large the data set is, while the mean is greatly affected by a value significantly different in magnitude from the other values. So, as a single measure of central tendency, the median is the better choice. However, for a variety of reasons, the mean is the measure that is most often used. We'll see in the next two sections how the measure of how the data is spread allows us to use the mean.

EXAMPLE

The seating capacity for the 30 stadiums in Major League Baseball is given. Determine the median and mean for this data.

Stadium Name	Capacity
Tropicana Field	31,042
Progressive Field	35,051
Marlins Park	36,742
Fenway Park	37,731
Kauffman Stadium	37,903
PNC Park	38,362
Target Field	38,885
Petco Park	40,209
Guaranteed Rate Field	40,615
SunTrust Park	41,149
Minute Maid Park	41,168
Wrigley Field	41,268
Comerica Park	41,299
Nationals Park	41,339
Miller Park	41,900

Stadium Name	Capacity
AT&T Park	41,915
Citi Field	41,922
Great American Ball Park	42,319
Citizens Bank Park	43,651
Angel Stadium of Anaheim	45,477
Busch Stadium	45,529
Oriole Park at Camden Yards	45,971
Oakland–Alameda County Coliseum	47,170
Yankee Stadium	47,422
Safeco Field	47,943
Globe Life Park in Arlington	48,114
Chase Field	48,686
Rogers Centre	49,282
Coors Field	50,398
Dodger Stadium	56,000

Source: https://en.wikipedia.org/wiki/List_of_current_Major_League_Baseball_stadiums

Since there are 30 data values, the median of the data will be the mean of the fifteenth and sixteenth pieces of data: $\frac{41,900 + 41,915}{2} = 41,907.5$. The mean of the data is $\mu = \frac{\sum_{i=1}^{30} stadium_i}{30} = 42,881.8$. The data given represents the population of the MLB stadiums, so the parameter μ is used.

Frequency tables are often used when the amount of data is large. There are two possible cases for us to consider. In the first case, data values are distinct. In the second case, the data values are stored in intervals.

Find the median and mean for the following test scores:

Score	Frequency
100	5
94	10
90	25
86	28
80	17
75	5

We need to know the number of pieces of data in order to find the median. The sum of the frequency column, 90, tells us the number of data values. The median will be the mean of the forty-fifth and forty-sixth pieces of data. The forty-fifth and forty-sixth data points are each 86. Therefore, the median is 88.

The mean for this data is

$$\frac{5(100)+10(94)+25(90)+28(86)+17(80)+5(75)}{90} = 87.03 \text{ (rounded to the nearest}$$

hundredth). Symbolically, the mean equals $\dfrac{\sum\limits_{i=1}^{n} f_i x_i}{\sum\limits_{i=1}^{n} f_i}$ where the term

$f_i x_i$ is the product of the data score (x) and the frequency of the score (f). The denominator is the sum of all the frequencies and thus represents all the data in the data set.

EXAMPLE

Find the median and mean (rounded to the nearest tenth) for the data showing the number of calories in a sample of foods offered at a fast-food restaurant.

Calories	Frequency		Calories	Frequency
5	4		300	5
10	5		310	3
140	6		320	6
150	2		330	7
160	2		340	5
180	2		380	6
210	4		400	6
220	4		440	7
230	3		520	4
270	3		540	3

Source: http://www.nutritionsheet.com/facts/restaurants/fast-food

The number of pieces of data is the sum of the frequencies, 87. The median is the forty-fourth piece of data: 320 calories. The mean of this sample is

$$\bar{x} = \frac{24{,}930}{87} = 286.6 \text{ calories.}$$

EXAMPLE

The mean for the set of data presented in the accompanying table is 22.25. Determine the value of n.

Score	Frequency
15	10
19	15
24	n
25	12
30	8

Compute the sum:

$$\frac{10(15) + 15(19) + n(24) + 12(25) + 8(30)}{n + 45}$$ and set this equal to 22.5.

▶ Multiply both sides of the equation by $n + 45$:

$$150 + 285 + 24n + 300 + 240 = 22.25(n + 45).$$

▶ Combine terms:

$$24n + 975 = 22.25n + 1{,}001.25$$

▶ Gather like terms:

$$1.75n = 26.25$$

▶ Divide by 1.75:

$$n = 15.$$

A requirement for data that is presented with interval format is that the intervals formed cannot overlap and the intervals must all be of the same width. An interval table that contains the intervals 90–100, 80–90, 70–80, etc. has a width of 10 and operates as the intervals $90 < x \leq 100$, $80 < x \leq 90$, $70 < x \leq 80$, etc. The score associated with each interval is the midpoint of the interval.

EXAMPLE

▶ Find the mean of the data displayed in the table.

Interval	Frequency
90–100	23
80–90	46
70–80	53
60–70	12
50–60	4

▶ Insert a column to enter the score assigned to each interval.

Interval	Score	Frequency
90–100	95	23
80–90	85	46
70–80	75	53
60–70	65	12
50–60	55	4

▶ Compute the mean (rounded to the nearest tenth):

$$\frac{23(95) + 46(85) + 53(75) + 12(65) + 4(55)}{23 + 46 + 53 + 12 + 4} = \frac{11{,}070}{138} = 80.2$$

EXERCISE 10.1

1. Find the median and the mean for the given data.

   12   23   34   19   28   61   45   39   17

2. Find the median and the mean for the given data.

   12   23   34   19   28   61   45   39   17   120

3. The income (in millions of dollars) for the 25 highest paid musicians is shown in the given table. Compute the median and mean for this data.

Name	Income	Name	Income	Capacity	Income
Sean Combs	130	Metallica	66.5	Kenny Chesney	42.5
Beyoncé	105	Garth Brooks	60	Luke Bryan	42
Drake	94	Elton John	60	Celine Dion	42
The Weeknd	92	Paul McCartney	54	Jay-Z	42
Coldplay	88	Red Hot Chili Peppers	54	Bruno Mars	39
Guns N' Roses	84	Jimmy Buffet	50.5	Tiësto	39
Justin Bieber	83.5	Calvin Harris	48.5	Jennifer Lopez	38
Bruce Springsteen	75	Taylor Swift	44	The Chainsmokers	38
Adele	69				

Data Source: https://www.cheatsheet.com/entertainment/the-highest-paid-musicians-of-2017.html/?a=viewall

4. The following data represents the 2017 salaries (rounded to the nearest million dollars) for NBA players. (The 15 players who earned less than $1 million are excluded from this table.) Compute the median and mean salaries based on this data.

Salary	Frequency	Salary	Frequency	Salary	Frequency
34	1	20	4	8	15
33	1	19	5	7	17
31	1	18	7	6	28
30	2	17	9	5	18
29	3	16	5	4	28
28	3	15	7	3	40
26	3	14	10	2	58
25	3	13	9	1	99
24	5	12	11		
23	7	11	11		
22	4	10	10		
21	3	9	8		

Data Source: http://www.espn.com/nba/salaries

5. Determine the value of n if the mean for this data is 44.5.

Score	Frequency
20	14
30	16
40	21
50	n
60	12
70	14

Measures of Dispersion

Let's take a look at the 2017 data for the NBA. There is a $4.47 million difference between the median and mean salaries. The scatter plot for this data clearly shows that it is skewed to the left (that is, there are more data values on the left than on the right).

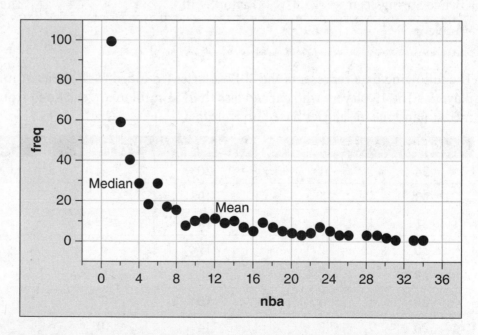

Half the salaries are less than or equal to $4 million (when values are rounded to the nearest million dollars). At first look, it seems that both values, though accurate, do not give a clear representation of the entire set of data. (The British Prime Minister, Benjamin Disraeli, is often cited for his quote "There are lies, there are damned lies, and there are statistics.")

To help clarify the nature of the data being summarized, the mean is usually reported with a second number representing how the data is dispersed (spread out). The simplest measure of dispersion is the **range** of the data. The range is the difference between the maximum and minimum values in the data set. For the NBA data, the range is $33 million.

Two more useful measures of dispersion (useful in that they are used in the branch of statistics called inferential statistics) are the **inter-quartile range (IQR)** and the **standard deviation**. The median represents the fiftieth percentile for a set of data. The first quartile (Q1) is the twenty-fifth percentile, and the third quartile (Q3) is the seventy-fifth percentile. The IQR is the difference of Q3 and Q1. (Q1 is midway between the minimum value and the median, while Q3 is midway between the median and the maximum value.) The IQR is used to determine whether a data value is "unusually" large or small. The rule of thumb is that a data value that is more than 1.5 times the IQR greater than Q3 is considered too large, while a number that is less than 1.5 times the IQR below Q1 is considered too small. Numbers that fit this definition of unusually too large or too small are called **outliers**.

The box and whisker plot is a graphical representation of data that displays the minimum, Q1, median, Q3, and maximum values of the data set. The five numbers are referred to as the 5 number summary for the data.

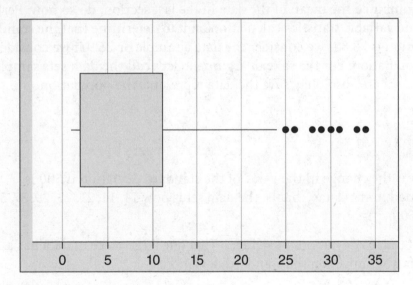

The minimum value is $1 million, the first quartile (Q1) = $2 million, the median = $4 million, the third quartile (Q3) = $12 million, and the maximum value is $34. There are 8 salaries (Stephen Curry, LeBron James, Paul Millsap, Blake Griffin, Gordon Hayward, Kyle Lowry, Mike Conley, and Russell Westbrook) whose salaries are so large that they are considered outliers. The IQR for this data is $10 million.

While the range and IQR give the reader a sense of spread about the data, the most commonly used measure of dispersion is the standard deviation. Roughly interpreted, the standard deviation gives the average difference between the data values and the mean of the data. (There is a bit more to it than that, but this interpretation should help give you a feel for the number.) The formulas for the parameter σ (lowercase sigma) and the statistics vary slightly.

$$\sigma = \sqrt{\sum_{i=1}^{n} \frac{\left(x_i - \overline{x}\right)^2}{n}} \text{ and } s = \sqrt{\sum_{i=1}^{n} \frac{\left(x_i - \overline{x}\right)^2}{n-1}}$$

The difference in the formulas is due to a concept called degrees of freedom and is something you will study when you take a course in statistics.

For data that is recorded with frequencies, the formulas are:

$$\sigma = \sqrt{\sum_{i=1}^{n} \frac{f_i\left(x_i - \overline{x}\right)^2}{n}} \text{ and } s = \sqrt{\sum_{i=1}^{n} \frac{f_i\left(x_i - \overline{x}\right)^2}{n-1}}$$

EXAMPLE

▶ Compute the standard deviation for the data: 14, 14, 21, 23, 29, 37, 53, 100.

▶ If you did not enter the data into a list on your graphing calculator when you computed the mean of the data in the last section, do so now. Perform the one variable statistics calculation on it to determine that the standard deviation is 28.8 if we consider the data a sample or 26.9 if we consider it a population. For the sake of argument, let's call the data sets samples unless we are absolutely sure the data represents the population.

EXAMPLE

▶ What is the change in the value of the standard deviation if 500 is included in the data? That is, the data set is now 14, 14, 21, 23, 29, 37, 53, 100, 500.

▶ The standard deviation is s = 156.9. That one large piece of data has a significant impact on the standard deviation.

EXAMPLE

Compute the standard deviation for the seating capacity of the Major League Baseball stadiums.

Stadium Name	Capacity	Stadium Name	Capacity
Tropicana Field	31,042	AT&T Park	41,915
Progressive Field	35,051	Citi Field	41,922
Marlins Park	36,742	Great American Ball Park	42,319
Fenway Park	37,731	Citizens Bank Park	43,651
Kauffman Stadium	37,903	Angel Stadium of Anaheim	45,477
PNC Park	38,362	Busch Stadium	45,529
Target Field	38,885	Oriole Park at Camden Yards	45,971
Petco Park	40,209	Oakland–Alameda County Coliseum	47,170
Guaranteed Rate Field	40,615	Yankee Stadium	47,422
SunTrust Park	41,149	Safeco Field	47,943
Minute Maid Park	41,168	Globe Life Park in Arlington	48,114
Wrigley Field	41,268	Chase Field	48,686
Comerica Park	41,299	Rogers Centre	49,282
Nationals Park	41,339	Coors Field	50,398
Miller Park	41,900	Dodger Stadium	56,000

These data represent all the major league stadia, so we are computing the parameter, rather than the statistic. The standard deviation is $\sigma = 5{,}082.3$.

Find the range, IQR, and standard deviation for the data showing the number of calories in a sample of foods offered at a fast-food restaurant.

Calories	Frequency	Calories	Frequency
5	4	300	5
10	5	310	3
140	6	320	6
150	2	330	7
160	2	340	5
180	2	380	6
210	4	400	6
220	4	440	7
230	3	520	4
270	3	540	3

If you have not already done so, this data should be recorded into two lists on your graphing calculator. This is still a one variable statistical calculation with the difference being the frequency of each value is not necessarily 1. The five number summary for the data is: min = 5, Q1 = 210, median = 320, Q3 = 380, max = 540. The range is 535 calories, and the IQR is 170 calories. The statement of the problem tells us this is a sample. The standard deviation is $s = 141.2$ calories.

EXERCISE 10.2

1. Find the range, IQR, and standard deviation for the given data.

    12   23   34   19   28   61   45   39   17

2. Find the range, IQR, and standard deviation for the given data.

    12   23   34   19   28   61   45   39   17   120

3. The income (in millions of dollars) for the 25 highest paid musicians are shown in the given table. Compute the IQR and standard deviation for this data.

Name	Income	Name	Income	Capacity	Income
Sean Combs	130	Metallica	66.5	Kenny Chesney	42.5
Beyoncé	105	Garth Brooks	60	Luke Bryan	42
Drake	94	Elton John	60	Celine Dion	42
The Weeknd	92	Paul McCartney	54	Jay-Z	42
Coldplay	88	Red Hot Chili Peppers	54	Bruno Mars	39
Guns N' Roses	84	Jimmy Buffet	50.5	Tiësto	39
Justin Bieber	83.5	Calvin Harris	48.5	Jennifer Lopez	38
Bruce Springsteen	75	Taylor Swift	44	The Chainsmokers	38
Adele	69				

Data Source: https://www.cheatsheet.com/entertainment/the-highest-paid-musicians-of-2017.html/?a=viewall

4. The following data represents the 2017 salaries (rounded to the nearest million dollars) for NBA players. Compute the range, IQR, and standard deviation salaries based on this data.

Salary	Frequency	Salary	Frequency	Salary	Frequency
34	1	20	4	8	15
33	1	19	5	7	17
31	1	18	7	6	28
30	2	17	9	5	18
29	3	16	5	4	28
28	3	15	7	3	40
26	3	14	10	2	58
25	3	13	9	1	99
24	5	12	11		
23	7	11	11		
22	4	10	10		
21	3	9	8		

Data Source: http://www.espn.com/nba/salaries

5. The number of calories in each of the Burger King sandwiches are displayed. Compute the median, mean, range, IQR, and standard deviation for the data.

220	350	460	610	770	930
260	370	460	630	770	970
260	380	490	640	790	1,000
300	390	510	670	790	1,010
310	400	520	670	800	1,070
320	420	520	690	830	1,090
320	450	530	690	850	1,160
330	450	570	750	850	1,250
340	460	590	760	920	1,310

Source: http://www.nutritionsheet.com/facts/restaurants/fast-food

Regressions

For the many years before calculators were used in the classroom, students had the perception that mathematics was not real because the answers to the equations done in class were always "nice" numbers (integers, terminating decimals, or fractions with reasonable denominators). The reality is that the entire world has to use numbers that are not always "nice." With the inclusion of technology in the classroom, particularly the graphing calculator, you have had the opportunity to see more and more real applications of mathematics. It is often the case that data is gathered from some research and, in analyzing the data, a mathematical equation is sought to relate two or more variables so that predictions can be made about future behavior.

EXAMPLE

A diver needs to be aware of the pressure that is being applied to his or her body and equipment when swimming in the ocean. The table below shows pressure data (in pounds per square inch—psi) for different depths (in feet).

Depth	Pressure (psi)
32.81	146.96
49.21	220.44
65.62	293.92
72.18	323.31
82.02	367.40
114.83	514.36

(a) Make a scatterplot of the data with depth as the independent variable.

(b) Determine the type of relationship that appears to be present between these variables.

(c) Determine the equation of best fit for the data.

(d) Predict the amount of pressure on a diver at a depth of 90 ft.

Solutions:

(a)

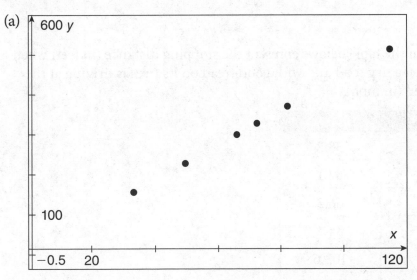

(b) The scatterplot indicates that there is a linear relationship between them.

(c) Use the Linear Regression tool of your calculator or computer program to determine the equation of best fit is pressure = 4.479 × depth – 0.003. The slope of this line is approximately 4.479 psi/ft while the vertical intercept is approximately 0 psi. The vertical intercept indicates that the pressure on the surface of the water is 0 psi, and the slope shows that the pressure will increase by 4.479 psi for every 1 foot of depth.

Regression equations can be used to predict an output value from an input value that is within the given data points but not from outside the given interval.

With an *r* value almost equal to 1 and a plot of the residuals showing a random distribution, we can be assured that the model is very good.

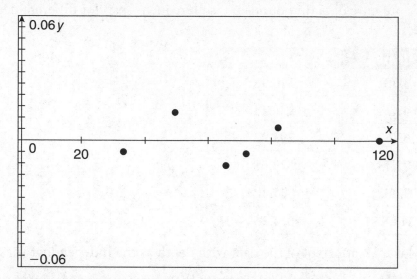

(d) According to this model, the pressure on a diver at a depth of 90 ft. will be 4.479 × 90 – 0.003 or 403.1 psi. (This calculation was used from the regression equation within the calculator and not the equation above which has the parameters rounded.)

EXAMPLE

The data in the table below represent the stopping distance (in feet) when a vehicle on a dry road and with good tread on its tires is driving at the given speed (in mph).

Speed	Distance
30	38.7
35	52.1
40	66.3
45	84.2
50	102.6
55	121.3
60	148.9
65	177.1
70	209.2
80	270.1

(a) Make a scatterplot for this data.

(b) Determine the type of relationship that appears to be present between these variables.

(c) Determine the equation of best fit for the data.

(d) Predict the stopping distance on a dry road with tires having good tread from a speed of 67 mph.

Solutions:

(a)

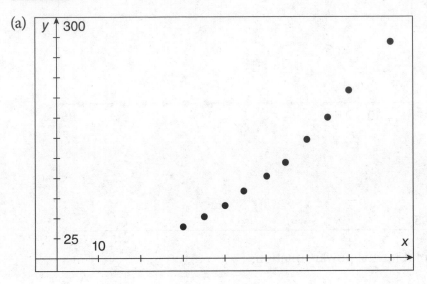

(b, c) The curvature of the graph might indicate that the relationship is exponential, it may be a power function, or it may be quadratic. A logarithmic relationship is unlikely because the graph is increasing. Use your calculator or computer to determine the equation of best fit using an exponential regression and a power regression.

	Equation	r value
Exponential	dist = 13.691(1.0400)^{speed}	1
Power	dist = 0.044$x^{1.987}$	0.99893
Quadratic	dist = 0.042x^2 − 0.004x + 0.55	–

The r value, called the Pearson correlation coefficient, gives a measure of the strength of the equation computer. It will always be the case that $-1 \leq r \leq 1$. The closer $|r|$ is to 1, the better the equation fits the data. (This rule only applies to linear, power, exponential, and logarithmic regressions. The reason for this is that each form of the curves can be linearized through the use of logarithms.) The exponential regression returns an r value of 1, while the power function gives an r value 0f 0.99893. The quadratic function does not return an r value.

If the plot of the residuals shows a distinct pattern, then the regression equation is not as good as it might be. The best regression equation will have a residual plot that is random.

▶ A way to compare models that cannot be linearized is to look at a residual plot. Residuals are the difference between the data value and the value predicted by the model. It is usually the case that the more random the residual plot, the better the model. In this case, however, the exponential model is a perfect fit because there are no residuals.

▶ Residual plot for the exponential model:

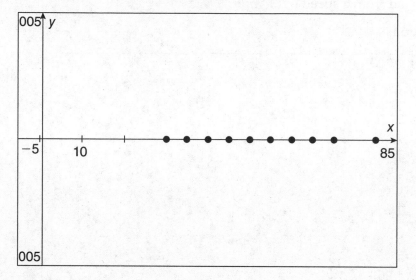

▶ Residual plot for the power function:

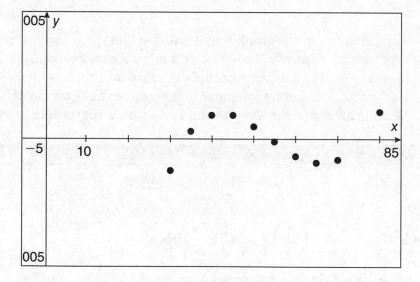

Residual plot for the quadratic model:

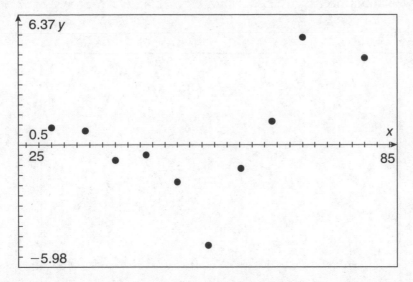

(d) The distance needed to stop a car traveling at 67 mph is dist = $13.691(1.0400)^{67} = 189.5$ ft.

EXAMPLE

A ball is dropped from a height of 20 feet and, as the ball bounces, the maximum height of each successive bounce is measured. Some of the data is not recorded. The data collected is shown in the accompanying table.

Bounce	Height (ft)
1	18.4
2	16.8
3	15.6
4	14.4
5	13.2
8	9.4
9	8.7

(a) Make a scatterplot for this data.

(b) Determine the type of relationship that appears to be present between these variables.

(c) Determine the equation of best fit for the data.

(d) Determine the height of the ball after the seventh bounce.

(e) Extrapolate from this data to predict the height of the ball after the twelfth bounce.

▶ Solutions:

(a)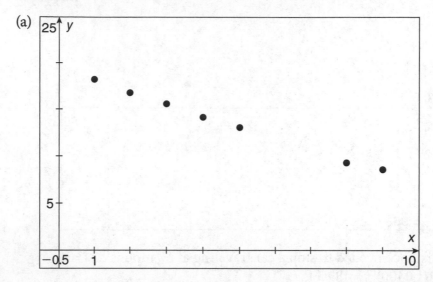

(b) The scatterplot indicates that the data fits an exponential model.

(c) Use your graphing calculator or computer to determine the equation of best fit is height = $20.6(0.9089)^{bounce}$ ($r = -0.9973$).

(d, e) The height of the ball after the seventh bounce will be $20.6(0.9089)^7 =$ 10.57 ft. and the height of the ball after the twelfth bounce is 6.56 ft.

EXAMPLE

▶ The table below gives the wind chill factors when the air temperature is 10°F.

Wind Speed (mph)	Wind Chill
10	−3.5
20	−8.9
30	−12.3
40	−14.8
50	−16.9
60	−18.6

(a) Make a scatterplot for this data.

(b) Determine the type of relationship that appears to be present between these variables.

(c) Determine the equation of best fit for the data.

(d) Predict the wind chill when the wind is blowing at 35 mph.

▶ Solutions:

(a)

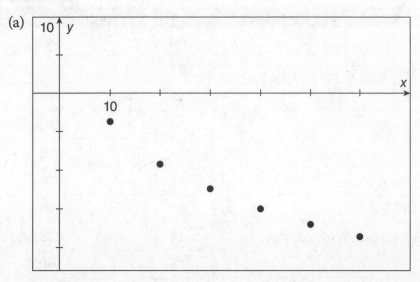

(b) The curvature of the scatterplot shows that a linear function is unlikely, and the presence of both positive and negative values eliminates exponential functions as a possibility. The scatterplot cannot represent a power function because it is impossible for a function of the form $f(x) = ax^b$ to produce both positive and negative outputs when the input values are always positive. Consequently, it appears that a logarithmic regression will best fit the data.

(c) The equation of best fit is wind chill = 16.12 – 8.42 ln(x) (The r value is –0.9992.)

(d) At a speed of 35 mph, the wind chill will be 16.12 – 8.42 ln(35) = –13.82°F.

EXERCISE 10.3

The data in the table below shows the number of grams of fat and the number of calories from fat in Burger King sandwiches.

Sandwich	Total Fat(g)	Calories from Fat
Whopper Sandwich	40	360
Whopper Sandwich Without Mayo	23	200
Whopper Sandwich with Cheese	48	430
Whopper Sandwich with Cheese without Mayo	30	270
Double Whopper Sandwich	58	520
Double Whopper Sandwich Without Mayo	41	370
Double Whopper Sandwich with Cheese	65	590
Double Whopper Sandwich with Cheese Without Mayo	48	430
Triple Whopper Sandwich	76	690
Triple Whopper Sandwich Without Mayo	59	530
Triple Whopper Sandwich with Cheese	84	760
Triple Whopper Sandwich with Cheese Without Mayo	66	600
Texas Whopper Sandwich	52	470
Texas Double Whopper Sandwich	70	630
Texas Triple Whopper Sandwich	88	790
Whopper Jr. Sandwich	20	180
Whopper Jr. Sandwich Without Mayo	11	100
Whopper Jr. Sandwich with Cheese	23	210
Whopper Jr. Sandwich with Cheese Without Mayo	15	130
Hamburger	11	100
Cheeseburger	15	130
Double Hamburger	19	170
Double Cheeseburger	27	240
Sourdough Bacon Cheeseburger	54	480
Rodeo Cheeseburger	18	160
Bacon Cheeseburger	17	150
Double Bacon Cheeseburger	30	270

Sandwich	Total Fat(g)	Calories from Fat
BK Double Stacker	37	330
BK Triple Stacker	51	460
BK Quad Stacker	65	580
A1 Steakhouse XT	61	550
Steakhouse XT	46	420
BK Burger Shots 2-pack	11	100
BK Burger Shots 6-pack	32	290
Tendergrill Chicken Sandwich	21	190
Tendergrill Chicken Sandwich without Mayo	9	80
Tendercrisp Chicken Sandwich	46	410
Tendercrisp Chicken Sandwich Without Mayo	22	200
Original Chicken Sandwich	39	350
Original Chicken Sandwich Without Mayo	16	140
Original Chicken Club Sandwich	43	390
Spicy Chick'n Crisp Sandwich	30	270
Spicy Chick'n Crisp Sandwich Without Mayo	12	110
BK Big Fish Sandwich	31	280
BK Big Fish Sandwich without Tartar Sauce	13	120
BK Veggie Burger	16	150
BK Veggie Burger with Cheese	20	180
BK Veggie Burger Without Mayo	7	70

Source: http://www.nutritionsheet.com/facts/restaurants/fast-food

1. Sketch a scatterplot for the number of calories from fat in a Burger King sandwich in terms of the number of grams of fat in the sandwich.

2. Determine the equation for the line of best fit for the number of calories from fat in a Burger King sandwich in terms of the number of grams of fat.

3. Explain the meaning of slope for this equation.

4. Predict the number of calories from fat in a Burger King sandwich if the sandwich contains 20 grams of fat.

An astronomical unit (au) is defined to be the average distance of the sun to the center of the earth. The table below contains the number of astronomical units the planets are from the sun and the number of Earth years it takes for the planet to make a revolution around the sun. Computations are based on Kepler's Third Law of planetary motion.

Planet	Distance from the Sun	Time for 1 Revolution Around the Sun
Mercury	0.39	0.24
Venus	0.72	0.62
Earth	1	1
Mars	1.52	1.88
Jupiter	5.2	11.86
Saturn	9.54	29.46
Uranus	19.18	84.01
Neptune	30.06	164.8

5. Make a scatterplot for this data.

6. Determine the type of relationship that appears to be present between these variables.

7. Determine the equation of best fit for the data.

8. Predict the time needed for the planetoid Ceres, which is 2.7 au from the sun, to make a complete revolution around the sun.

The table below gives the wind chill factors when the air temperature is 0°F.

Wind Speed (mph)	Wind Chill
10	−15.9
20	−22
30	−25.9
40	−28.8
50	−31.1
60	−33.1

9. Make a scatterplot for this data.

10. Determine the type of relationship that appears to be present between these variables.

11. Determine the equation of best fit for the data.

12. Predict the wind chill when the wind is blowing at 45 mph when the air temperature is 0°F.

Normal Distribution

You should be familiar with the notion that in almost every endeavor there is a continuum of outcomes from the completely unacceptable to the truly exceptional and that the graphical representation of this continuum is known as the bell curve.

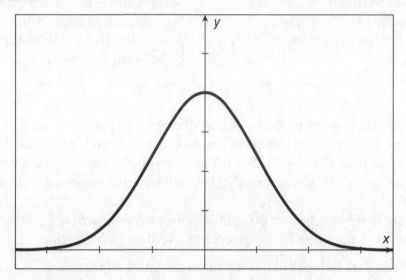

The probability for a continuous random variable (a variable that is measured rather than counted) is computed by measuring the area under the bell curve. (This was originally done with calculus, then with a table of values, but is now done with technology available to you.) To compute the area under the bell curve requires the identification of two points—a left endpoint and a right endpoint—needed to define the area in question. What if the left endpoint and the right endpoint are the same number? The result will be a line segment from the x-axis to the point on the graph of the bell curve. Since a line segment is a 1-dimensional figure, it has no area. This illustrates an important aspect of the continuous random variable—probabilities are computed for an interval, however small the interval needs to be, but not a point. That is, we can calculate the probability for a continuous random variable that x is between 1.5 and 2.5, between 1.7 and 2.3, between 1.999 and 2.001, but never at $x = 2$.

EXAMPLE

▶ The height of students at Fort Mill High School is normally distributed with a mean of 69.5 inches and a standard deviation of 2.8 inches. If a student is selected at random, what is the probability that the student's height will be between 63.9 and 75.1 inches?

▶ The diameter of ball bearings made under a particular process is normally distributed with a mean of 1.2 cm with a standard deviation of 0.05 cm. If a ball bearing made under this process is selected at random, what is the probability that its diameter will be between 1.1 and 1.3 cm?

Believe it or not, these two questions have the same answer. Reread the questions and observe that the lower bound for each region is two standard deviations less than the mean while the upper bound is two standard deviations greater than the mean. Consequently, they will have the same area under the bell curve.

There are three benchmark probabilities for normal distributions that are considered to be common knowledge among students of statistics:

- 68% of the data lies within one standard deviation of the mean
- 95% of the data lies within two standard deviations of the mean
- 99.7% of the data lies within three standard deviations of the mean

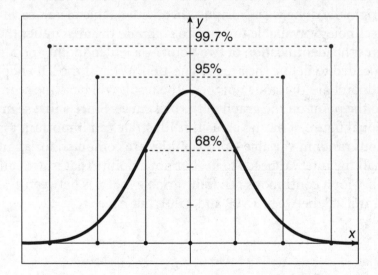

EXAMPLE

▶ Let x be the variable of a normally distributed quantity with mean, μ, and standard deviation, σ. What is the probability that $P(x > \mu)$? (Translation: What is the probability that x is greater than the mean?)

▶ When you examine the area under the normal curve in the diagram above, you see that it is symmetric about the mean, μ. Since the total area under the curve is 1, half the area must be to the right of μ and half to the left of μ. Therefore, $P(x > \mu) = 0.5$.

EXAMPLE

▶ The height of students at Fort Mill High School is normally distributed with a mean of 69.5 inches and a standard deviation of 2.8 inches. If a student is selected at random, use the benchmark probabilities that the student's height will be:

(a) between 66.7 and 72.3 inches

(b) between 69.5 and 75.1 inches

(c) over 72.3 inches

(d) under 75.1 inches

(e) under 66.7 inches or over 72.3 inches

▶ Solutions: Let h represent he height of a randomly selected student.

(a) $P(66.7 < h < 72.3) = 0.68$ because these numbers are each 1 standard deviation from the mean.

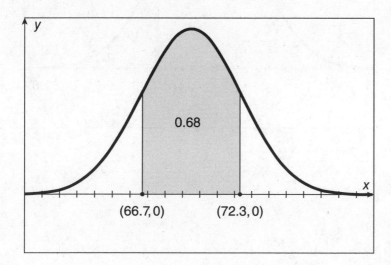

(b) $P(69.5 < h < 75.1) = 0.475$ because this the region from the mean to the point 2 standard deviations above the mean.

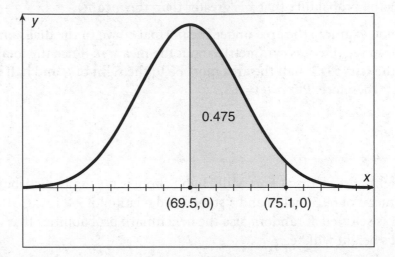

(c) $P(h > 72.3)$ represents the region to the right of 1 standard deviation above the mean. Knowing that the area to the right of the mean is 0.5 and the area between the mean and 1 standard deviation above the mean is 0.34, $P(h > 72.3) = 0.5 - P(69.5 < h < 72.3) = 0.16$.

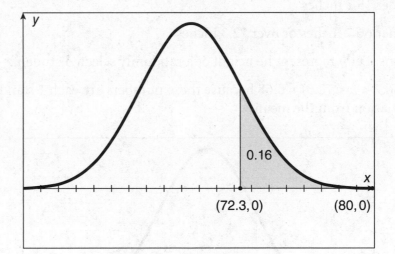

(d) $P(h < 75.1)$ is the area to the left of 75.1. Since 75.1 represents the point 2 standard deviations above the mean, use $P(h < 75.1) = P(h < 69.5) + P(69.5 < h < 75.1) = 0.5 + 0.475 = 0.975$.

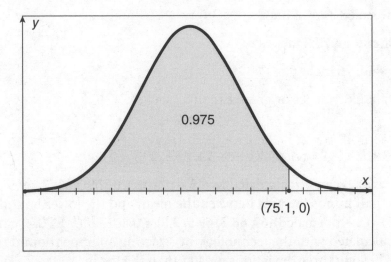

(e) $P(h < 66.7$ or $h > 72.3)$ is the complement of $P(66.7 < h < 72.3)$. $P(h < 66.7$ or $h > 72.3) = 1 - P(66.7 < h < 72.3) = 1 - 0.68 = 0.32$.

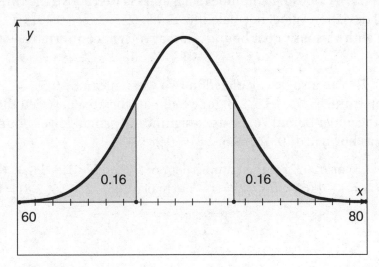

Most graphing calculators, spreadsheet programs, and computer programs with CAS capability have built-in functions that can compute probabilities. The TI 83/84 series and the Nspire calculators have a normcdf function that can compute probabilities for all values of the input variable. The parameters for this function are normcdf(lower bound, upper bound, mean, standard deviation).

Using the distribution of heights of students at Fort Mill High School, determine the probability that the height of a randomly selected student is:

(a) between 68 and 74 inches

(b) greater than 70.2 inches

(c) less than 76 inches

(d) less than 65 inches or greater than 72 inches

Solutions:

(a) $P(68 < h < 74)$ = normalcdf(68, 74, 69.5, 2.8) = 0.6499.

(b) $P(h > 70.2)$ can be calculated as 0.5 – normalcdf(69.5, 70.2, 69.5, 2.8) (one half minus the area between the mean and the lower bound, 69.5). 0.5 – normalcdf(67, 68.3, 69.5, 2.8) = 0.4013. $P(h > 70.2)$ can also be calculated with the technology by selecting an upper bound that is well beyond three deviations greater than the mean.

For this example, three deviations larger than the mean, 69.5, is 69.5 + 3(2.8) = 77.9. Choosing a number such as 85 is well above the three standard deviation value. normalcdf(69.5, 85, 69.5, 2.8) = 0.4013. (Try working with various upper bounds so that you get comfortable with this notion.)

(c) $P(h < 76)$ can also be calculated in two ways. It can be 0.5 + normalcdf(69.5, 76, 69.5, 2.8) (one half plus the area between the mean and the upper bound 76) or use a significantly small lower bound, such as normalcdf(0, 76, 69.5, 2.8) = 0.9899.

(d) $P(h < 65$ or $h > 72)$ is the complement of $P(65 < h < 72)$. $P(h < 65$ or $h > 72) = 1 – P(65 < h < 72) = 1 –$ normalcdf(65, 72, 69.5, 2.8) = 0.2400.

EXERCISE 10.4

The heights of the female students in the senior class at Providence High School are normally distributed with a mean 64 inches and a standard deviation of 2.6 inches. Use this data and to answer questions 1–6.

1. Compute the probability that the height of a randomly selected senior is between 61 and 63.5 inches.

2. Compute the probability that the height of a randomly selected senior is between 56 and 66.5 inches.

3. Compute the probability that the height of a randomly selected senior is greater than 63.5 inches.

4. Compute the probability that the height of a randomly selected senior is less than 66.5 inches.

5. Compute the probability that the height of a randomly selected senior is less than 56 inches or greater than 66 inches.

6. Compute the probability that the height of a randomly selected senior is less than 58.5 inches or greater than 69 inches.

A soda dispenser dispenses soda with a normal distribution having a mean of 15.87 fluid ounces with a standard deviation of 0.15 ounces. A cup of soda is randomly selected from all the cups of soda that have been dispensed. Use this information to answer questions 7–10.

7. What is the probability that the number of fluid ounces dispensed is between 15.78 and 16.2?

8. What is the probability that the number of fluid ounces dispensed is less than 16.1?

9. What is the probability that the number of fluid ounces dispensed is more than 16.4?

10. What is the probability that the number of fluid ounces dispensed is less than 15.85 or more than 16.2?

11. Waiting time for a teller at the Eastside Federal Credit Union on a Friday night is normally distributed with a mean of 4.3 minutes and a standard deviation of 0.25 minutes. The bank manager is concerned with customer satisfaction and has initiated a policy of giving a $5 gift card to a customer who has to wait longer than 4.75 minutes to be served. What is the probability that a Friday night customer will receive a $5 gift card from the manager?

CHAPTER 11

Inferential Statistics

Describing the center and/or the spread of a set of data are interesting enough but not of much use unless we intend to do something with them such as debate which player or entertainer had the better season, made more money, or some such argument. The strength of gathering data is in the ability to use it to make predictions about population parameters in order to plan for future activities or to test the validity of claims made by entities such as businesses, governments, or researchers. While the study of inferential statistics covers a wide variety of scenarios, we will limit our discussion to large sample sizes.

One of the techniques used to gather data is simulation, a technique using probability to model the sampling process. The importance of this is that simulation is faster and cheaper. Imagine, for example, that you want to test the breaking strength of a rivet or shatterproof glass. Since it costs money to break these items and then they cannot be used again, probabilistic models are frequently employed.

Basic Concepts

Randomization is a major issue when it comes to the study of probability and inferential statistics (the application of probability). Designing surveys and sampling processes is tricky work because one wants to avoid any type of bias when collecting data. For example, if members of the high school prom committee want to determine the type of music those attending the prom would want to have, asking the members of only one classroom might not be very indicative of the entire group. Notice that it is not necessarily biased but the process does leave open that possibility.

EXAMPLE

▶ The members of the high school prom committee want to determine the type of music that those attending the prom would want to have. State three ways that the committee can randomly select students to survey.

1. The names of all the school's students can be accessed through a database. Use a random number generator to pick names based on their position within the list.

2. Separate the students by their grade level and then randomly choose equal numbers of students from each grade level.

3. Randomly select a set of homerooms from a list of all the homerooms in the school and ask the students in each homeroom to respond to the survey.

▶ These are three possible solutions. There are many others that can be used.

EXERCISE 11.1

Follow the instructions for each of the problems below.

1. A polling service is trying to predict the outcome of an upcoming national election. It randomly selects 1,000 people who live in the northeastern part of the United States. Does this represent an unbiased survey? Explain.

2. A polling service is trying to predict the outcome of an upcoming national election. Describe two processes that could be used to collect an unbiased set of data.

Central Limit Theorem and Standard Error

As we saw in Chapter 10, the normal distribution enables us to compute the probability that a piece of data will lie within an interval provided we know the mean and standard deviation of the distribution. While not all distributions are normal (think of a single fair die, the probability for each outcome 1 through 6 is the same, one-sixth, making the distribution uniform), there is a theorem in statistics that allows us to apply the normal distribution while attempting to make inferences about the center of the distribution.

Repeated random samples are taken from a population and the mean of each sample is recorded. The **Central Limit Theorem** states that as the sample size gets larger, the distribution of the sample mean can be approximated by the normal distribution. This is true whether or not the original distribution is normal. That is to say, if we take enough repeated samples from a population, compute the mean of the sample, and analyze the distribution of these means, the distribution will be approximately normal provided that the size of each sample is sufficiently large.

Two key pieces to this theorem are given a population with mean μ and a standard deviation σ:

- The mean of all the sample means, $\mu_{\bar{x}}$, is the same value.
- The standard deviation of the means (called the **standard error of the means**) $\sigma_{\bar{x}} = \dfrac{\sigma}{\sqrt{n}}$.

We can then apply the normal distribution to compute probabilities for the mean of the selected sample. This is the basis for the work on inferential statistics involving the parameter's mean and proportions and large samples. (There is a great deal more to the topic of inferential statistics that is well beyond the scope of this course. The intent at this level is to give you, the student, a feel for how decisions are made based on the use of statistics.)

> A key piece of the Central Limit Theorem is that if the sample sizes are large enough, the means of the samples will be (approximately) normally distributed, no matter what the distribution of the original data might be.

EXAMPLE

> A large number of samples of size 50 are drawn from a population with mean 73 and standard deviation of 2.3. What is the mean and standard error for these sample means?

> The mean remains as 73 but the standard error of the mean is $\dfrac{2.3}{\sqrt{50}} = 0.325$.

▶ A random sample of size 50 is drawn from a population with mean 73 and standard deviation of 2.3. What is the probability that the mean of the selected sample is between 72.5 and 73.9?

▶ Using the results of the previous problem, the standard error of the mean is $\dfrac{2.3}{\sqrt{50}}$. The probability that the mean is between 72.5 and 73.9 is

$$P\left(72.5 < \overline{x} < 73.9\right) = \text{normCDF}\left(72.5, 73.9, 73, \dfrac{2.3}{\sqrt{50}}\right) = 0.9350. \text{ (Use your}$$

graphing calculator to find this value.)

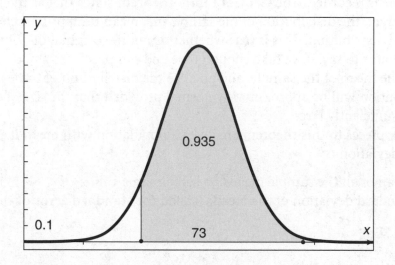

▶ A random sample of size 36 is drawn from a population that is normally distributed with a mean of 50 and a standard deviation of 4.8. What is the probability that the mean of the sample is greater than 52?

▶ The mean of the distribution of sample means is 50 and the standard

deviation is $\dfrac{4.8}{\sqrt{36}} = 0.8$. The probability that the sample mean is greater

than 52 is $P\left(\overline{x} > 52\right) = 1 - P\left(0 < \overline{x} < 52\right) = 0.0062.$

The sampling distribution for proportions, p, is similar to that of the mean. The distribution of the sample proportions can be approximated by a normal distribution with sample size n if the product $np \geq 10$ and $n(1 - p) \geq 10$. The

mean of the proportions will be p and the standard error is equal to $\sqrt{\dfrac{p(1-p)}{n}}$.

EXAMPLE

Repeated samples of size 64 are drawn from a population for which 75% are in favor of raising the gasoline tax for the purpose of gaining revenue to improve road conditions. What is the mean and standard error for these samples?

The mean proportion for the samples is that of the population, 0.75, while the standard error for the proportion is $\sqrt{\dfrac{(0.75)(0.25)}{64}} = 0.054$.

EXAMPLE

A sample of size 64 is drawn from a population for which 75% are in favor of raising the gasoline tax for the purpose of gaining revenue to improve road conditions. What is the probability the proportion of those selected in favor of such legislation is greater than 70%?

We can assume that the distribution of the proportions is normal with proportion 0.75 and standard error 0.054. We have $(0.75)(64) = 48$ and $(64)(.25) = 16$, thus meeting the criteria for assuming a normal distribution. The probability that more than 70% of the sample voters favor the legislation is:

$$P(p > 0.70) = 1 - P(0 < p < 0.70) =$$

$$1 - \text{normCDF}\left(0, 0.7, 0.75, \sqrt{\frac{(.75)(.25)}{64}}\right) = 0.822.$$

EXAMPLE

A sample of size 25 is drawn from a population for which 60% are in favor of raising the gasoline tax for the purpose of gaining revenue to improve road conditions. What is the probability the proportion of those selected in favor of such legislation is less than 70%?

We know from the last problem that the conditions of the problem allow us to assume the distributions of the sample proportions will be normal. The probability that less than 70% of the sample favor the legislation:

$$P(p < 0.7) = 0.5 + P(0.6 < p < 0.7) = 0.5 + 0.346 = 0.846.$$

EXAMPLE

▶ A sample of size 100 is drawn from a population from which 20% oppose tax reform. What is the probability that less than 15% of the sample oppose tax reform?

▶ The distribution of the sample proportions are normally distributed because $(100)(0.20) = 20$ and $(100)(0.80) = 80$ are both in excess of 10. The mean proportion for the sampling distribution is 0.20 while the standard error is $\sqrt{\dfrac{(0.2)(0.8)}{100}} = 0.04$. Therefore,

$$P(p < 0.15) = 0.5 - P(0.15 < p < 0.2) = 0.106.$$

EXAMPLE

▶ A sample of size 1,000 is drawn from a population from which 20% oppose tax reform. What is the probability that less than 15% of the sample oppose tax reform?

▶ The distribution of the sample proportions are normally distributed because $(1,000)(0.20) = 200$ and $(1,000)(0.80) = 800$ are both in excess of 10. The mean proportion for the sampling distribution is 0.20 while the standard error is $\sqrt{\dfrac{(0.2)(0.8)}{1000}} = 0.013$. Therefore,

$$P(p < 0.15) = 0.5 - P(0.15 < p < 0.2) = 0.000039.$$

▶ The size of the sample has a significant impact on the result because the distribution is much narrower.

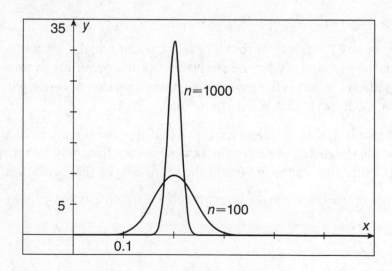

EXERCISE 11.2

For questions 1–4, determine the standard error for the parameter indicated in each problem.

1. Standard deviation = 7.5; sample size = 49

2. Standard deviation = 15.3; sample size = 50

3. Proportion = 0.7; sample size = 150

4. Proportion = 0.5; sample size = 500

The height of the students at Providence High School has a mean of 69.3 inches with a standard deviation of 4.3 inches. A random sample of 50 students is selected and their heights measured.

5. What is the probability that the mean height of the students is between 68.5 and 69.3 inches?

6. What is the probability that the mean height of the students is less than 68 inches?

7. What is the probability that the mean height of the students is greater than 70 inches?

Ninety-nine percent (99%) of all the batteries made at the Pineville factory meet the manufacturer's specifications. A random sample of 400 batteries is selected for testing.

8. What is the probability that between 98.5% and 99.5% of the batteries meet the manufacturer's specifications?

9. What is the probability that at least 99.5% of the batteries meet the manufacturer's specifications?

10. What is the probability that less than 0.5% of the batteries **fail** to meet the manufacturer's specifications?

Standardized (z) Scores

You read in the section on the Normal Distribution in Chapter 11:

▶ The height of students at Fort Mill High School is normally distributed with a mean of 69.5 inches and a standard deviation of 2.8 inches. If a student is selected at random, what is the probability that the student's height will be between 63.9 and 75.1 inches?

▶ The diameter of ball bearings made under a particular process is normally distributed with a mean of 1.2 cm with a standard deviation of 0.05 cm. If a ball bearing made under this process is selected at random, what is the probability that its diameter will be between 1.1 and 1.3 cm?

▶ Believe it or not, these two questions have the same answer. Reread the questions and observe that the lower bound for each region is two standard deviations less than the mean while the upper bound is two standard deviations greater than the mean. Consequently, they will have the same area under the bell curve.

A process used for comparing raw values from different distributions is to determine the number of standard deviations from the mean a data point is located. This value is called a **standardized score** and, as it is traditionally represented by the variable z, is also called a **z-score**. (Before the availability of computing devices to compute probabilities under the normal curve, the values were computed from a table of the standard normal distribution. This distribution had a mean of 0 and a standard deviation of 1.)

▶ Jack, Beth, and Juan, three friends who are each in different sections of a course in Algebra II, meet after their exams in the statistics units are returned and are surprised to see that they each scored 90% on the test. Jack commented that he was really pleased with his result because the average grade on the exam in his class was 87. Beth also expressed her pleasure at her score because the average grade in her class was 88. Juan seemed a little upset by this conversation. He explained that the average score in his class was 93. (The teacher confirmed that all three versions were comparable in their level of difficulty and that the scores on all three exams had a standard deviation of 2 points.) As people will often do, the three friends discuss which of them did better. Who had the "better" score?

Jack argued that he did the best since his score was 1.5 standard deviations higher than the mean while Beth scored 1 standard deviation above the mean and Juan scored 1.5 standard deviations below the mean. (One might also argue that scoring 90% on a statistics exam is quite an accomplishment, but that does not lead to an efficient ranking system.)

You know that the probability a data value from a variable that is normally distributed will lie within one standard deviation of the mean is approximately 68%. Said differently, $P(-1 < z < 1) = 68\%$. (Notice that the phrase "within one standard deviation" means the interval from one standard deviation below the mean to one standard deviation above the mean.)

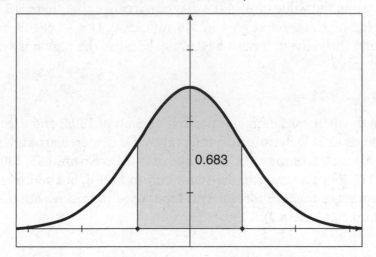

Between which two standard scores, symmetric about the mean, will 50% of the data lie? Under the old system, you would search the table looking for a z-score associated with a probability of 0.25 (only probabilities for z-scores greater than 0 were given because of the symmetry of the normal curve) and the negative of that score, which would provide the endpoints of the required interval. The trouble was that one could not find 0.25 in the table of probabilities and had to interpolate to get the value needed.

Rather than go through that extra work, we will take advantage of the invNorm function that is available on your calculator. The structure of the command for this function is invNorm(Area, Mean, Standard Deviation). It is important to note that the Area input represents the area under the normal curve to the **left** of the value we seek.

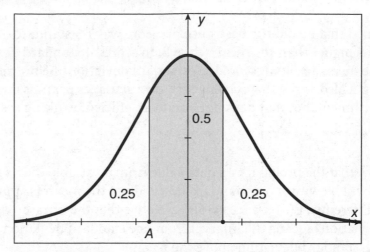

To determine the value of A in the diagram, realize that there is 25% of the area to the left of A. Therefore, A = invNorm(0.25, 0, 1) = −0.67449. You can easily tell that the value of B must be 0.67449 because the region is symmetric about the mean, 0.

EXAMPLE

▶ Suppose the distribution in question has a mean of 130.2 and a standard deviation of 12.3. Determine the interval that is centered around the mean and contains 50% of the data. The value of A is invNorm(0.25, 130.2, 12.3) = 121.904. You can still use symmetry to find B, but you might it easier to replace the 0.25 for the Area input used to find A with 0.75. B = invNorm(0.75,149.3,18.2) = 138.496.

EXAMPLE

▶ Problem: A variable is normally distributed with a mean of 547.3 and a standard deviation of 37.9. Between which two values, symmetric about the mean, will 90% of the data be located?

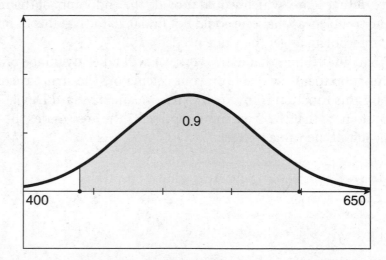

▶ Solution: A = invNorm(0.05, 547.3, 37.9) = 484.96 and B = invNorm(0.95, 547.3, 37.9) = 609.64.

EXAMPLE

Problem: The heights of the students at Providence High School has a mean of 69.3 inches with a standard deviation of 4.3 inches. What is the height that represents the point where only 2.5% of the students are taller?

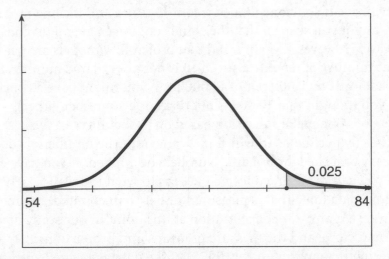

Solution: If 2.5% of the area is to the right of point X, then 97.5% must be to the left. Therefore, X = invNorm(0.975, 69.3, 4.3) = 77.73 inches.

EXERCISE 11.3

For problems 1–3, determine the z-score for the given value of x.

1. $x = 35$, mean = 25, standard deviation = 4

2. $x = 35$, mean = 28, standard deviation = 4

3. $x = 35$, mean = 31, standard deviation = 2.5

For problems 4 and 5, determine the values of A and B assuming that they are symmetric about the mean.

4. Mean = 78.2, standard deviation = 6.5, area between A and B is 80%.

5. Mean = 940, standard deviation = 73.9, area between A and B is 60%.

6. The volume of soft drink in a 1 liter bottle of Monahan's Famous Fluids is normally distributed with a mean of 1.02 liters and a standard deviation of 0.01 liters. Ninety-nine (99) percent of the bottles contain more than X liters. What is the value of X?

7. Mrs. Netoskie rides hunters in equestrian competitions. The heights of these hunters are normally distributed with a mean of 14 hands and a standard deviation of 1.1 hands. What is the height at which only 4% of the horses are taller?

Inferential Statistics

As was noted earlier in the chapter, statistical decisions are based on the study of probability. A fair coin is one in which the chances of getting heads or tails are equally likely. If someone flipped a coin 10 times and it came up heads each time, might you suspect that the coin is not fair? One might consider this to be the case. However, keep in mind that probability models are based on a very large number of repetitions, and 10 is not a very large number. If the coin showed heads on 9,000 out of 10,000 flips, you might have a concern (and hopefully you thought that 10,000 is not that large a number either).

So, how does one make a decision based on probability? In the case of discrete cases (experiments in which one can count the number of successes), it is necessary that the Law of Large Numbers be applied. If you have enough examples, you can make an educated guess as to what the reality is. In the case of continuous data (data that is measured rather than counted), we rely on the normal curve. (Again, our concentration at this point in our study of statistics is on the normal curve and the measure of means and proportions.)

Considering that approximately 99.7% of the data associated with a variable that is normally distributed will lie within three standard deviations from the mean, it is safe to assume that statistics that fall in the interval do so because of natural variability rather than because of an anomaly. It is important to note that you are never 100% certain that the decision made based on the results of a sample is accurate. The best one can hope for is to minimize the chance that the decision is incorrect.

> Not all samples drawn from the same population will have the same mean. Understanding that, we allow for values that differ from the mean so long as they are not too far from the mean. This is the basic rule used in determining confidence intervals and in performing tests of hypotheses.

CONFIDENCE INTERVALS

Let's begin with the situation that the population parameter for a continuous variable is unknown and the purpose of the sample is to get an estimate of its value. Using the Central Limit Theorem, we know that so long as we choose a sample large enough (curiously, 30 turns out to be the magic number for a sufficiently large sample size), the distribution of the corresponding statistic (mean or proportion) will be normally distributed. (You'll find that as you study more about statistics that authors will start with the case that we assume we do not know the population mean but do know the population standard deviation. That seems like a bit of a reach, so we'll work with the case that neither is known but will use the sample standard deviation in our calculations.) We acknowledge that our response is an estimate by giving a measure of the level of confidence we have with our result (and hence the term **confidence interval**).

EXAMPLE

A random sample of 50 banking customers who were at a particular branch during the lunch hour is selected. The time they waited while in line before a teller helped them is measured. The mean of the data is 5.7 minutes with a standard deviation of 0.6 minutes. Determine an interval for the mean waiting time of a lunchtime customer at this branch using a 95% level of confidence.

We begin by looking at a standard normal curve (remember, the mean is 0 and the standard deviation is 1). What two values will 95% of the data lie between?

$$A = \text{invNorm}(.025,0,1) = -1.96 \text{ while } B = \text{invNorm}(.975,0,1) = 1.96.$$

We know that z-scores are computed with the formula $z = \dfrac{x - \mu}{\sigma}$. In this case, the data value is the mean of the sample and the spread is actually the standard error of the mean. Since we are using the sample standard deviation, the equation is $z = \dfrac{\overline{x} - \mu}{\dfrac{s}{\sqrt{n}}}$. Solve this equation for μ,

$\mu = \overline{x} - z\left(\dfrac{s}{\sqrt{n}}\right)$. Because the z-score will always be ± the same value, the formula for computing the endpoints of the confidence interval becomes $\mu = \overline{x} \pm z\left(\dfrac{s}{\sqrt{n}}\right)$.

In the case of the waiting time for the customers at the bank, the mean waiting time is $5.7 - (1.96)\left(\dfrac{0.6}{\sqrt{50}}\right) = 5.534$ minutes to $5.7 + (1.96)\left(\dfrac{0.6}{\sqrt{50}}\right) = 5.866$ minutes.

Fortunately, we can use the invNorm command to get these same results much more quickly. Rather than use 0 as the mean and 1 as the standard deviation, use the mean and standard deviation from the problem. Lower bound is $\text{invNorm}\left(0.025, 5.7, \dfrac{0.6}{\sqrt{50}}\right) = 5.534$ minutes and the upper bound is $\text{invNorm}\left(0.975, 5.7, \dfrac{0.6}{\sqrt{50}}\right) = 5.866$ minutes.

EXAMPLE

▶ A third way to determine the endpoints of the confidence interval is to use the statistics tools on your graphing calculator. Choose STATS, Confidence Intervals, and z-interval. In this example, we have the mean and standard deviation for the sample, so choose Stats rather than Data for the input. Enter the standard deviation of the sample, the mean of the sample, the size of the sample, and the level of confidence. (The default value for the level of confidence is 95%.) Once entered, press OK to compute.

zInterval 0.6,5.7,50,0.95: *stat.results*	"Title"	"z Interval"
	"CLower"	5.53369
	"CUpper"	5.86631
	"x̄"	5.7
	"ME"	0.166308
	"n"	50.
	"σ"	0.6

▶ The lower bound, CLower, and upper bound, CUpper, are displayed along with the mean, mean error, sample size, and standard deviation. (The result, ME (mean error) is half the difference between the upper and lower bounds.)

EXAMPLE

▶ Problem: A random sample of 40 students from Hilltop High School had their heights measured. The average for the sample was 67.8 inches with a standard deviation of 3.4 inches. Determine an interval for the true mean height of the students at Hilltop High using a 90% level of confidence.

▶ Enter the data into your confidence interval tool to get the results

zInterval 3.4,67.8,40,0.9: *stat.results*	"Title"	"z Interval"
	"CLower"	66.9157
	"CUpper"	68.6843
	"x̄"	67.8
	"ME"	0.884252
	"n"	40.
	"σ"	3.4

▶ We are 90% confident that the mean height of the students at Hilltop High School is between 66.92 inches and 68.68 inches.

▶ Using the TI-84, press the Stat key, slide right to tests, choose option 7 z-Interval, enter the values, press Enter when the cursor is on calculate.

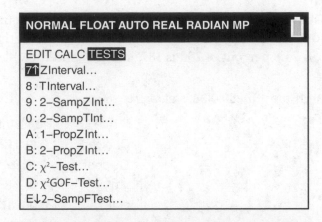

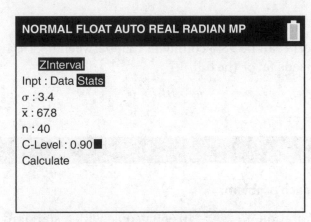

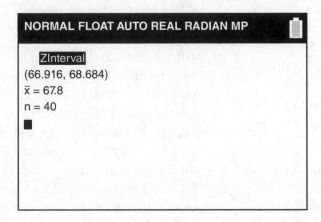

EXAMPLE

> A survey of 250 citizens of voting age is taken in a town to determine if there is support for a bill proposing to raise the state tax by 0.5% to be used to cover the cost for recycling, and 74.8% of those surveyed indicate that they support the bill. Determine an interval for the proportion of the population that support the bill at the 98% level of confidence.

> Use the statistics tool on your calculator to determine the bounds. This technique requires that you give the number of successes (rather than the percentage). So 74.8 percent of 250 is 187, n is 250, and the C-level is 0.98.

zInterval_1Prop 187,250,0.98: *stat.results*

"Title"	"1–Prop z Interval"
"CLower"	0.684121
"CUpper"	0.811879
"$\hat{p}$"	0.748
"ME"	0.063879
"n"	250.

> The town officers can be 98% confident that between 68.4% and 81.2% of the voting citizens favor the bill.

EXERCISE 11.4

Determine the confidence intervals for each problem.

1. The personnel department of a corporation wants to estimate the average amount of money spent annually on medical expenses. The results of a random sample of 50 employees show an average annual expense of $752.18 with a standard deviation of $90.30. Determine a 95% confidence interval for the average amount of money spent annually on medical expenses.

2. The manager of a rock and roll band wants to estimate the mean number of people who attend a concert put on by her clients. A random sample of the band's last 30 concerts shows an average attendance of 10,650 people with a standard deviation of 1,190. Determine an interval for the average attendance at the 90% level of confidence.

3. An automobile dealership manager wants to determine the proportion of new car transactions that have the customer select a lease option rather than purchase. The manager randomly selects 80 monthly records and determines that 55% of all transactions involve a lease option. Determine an interval for the proportion of monthly transactions on new cars that involve a lease option at the 98% level of confidence.

HYPOTHESIS TESTS

Your friend claims, "The average number of minutes of ad time on this radio station each hour is 20 minutes."

You respond, "I disagree."

What are the implications of the statement "I disagree"? Do you believe that the station has less than 20 minutes of ad time per hour, more than 20 minutes per hour, or just that it is not 20 minutes of ad time per hour on average? When performing a statistical test of a hypothesis, it is important to know which condition you are claiming as an alternative. Here's why.

EXAMPLE

▶ Without being able to examine the entire population under consideration, there will always be the possibility that the claim being made is accurate and the alternative is incorrect. The goal of the test of hypothesis is to make the probability that this happens is as small as is reasonably acceptable. A classic argument comes from our legal system. The charge is that the local manufacturing plant is polluting the town's water supply. The operating tenet in the legal system is "Innocent until proven guilty." The prosecutors present their evidence in an attempt to show beyond reasonable doubt that the company is guilty. The defense lawyer, if unable to out-and-out prove the innocence of the defendant, tries to present evidence to show reasonable doubt. When both sides are done, the case goes to the jury for deliberation. Due to the premise of innocent until proven guilty, the initial hypothesis is that the company is innocent. There are four possible scenarios.

A: The company is innocent and the jury finds it guilty. An incorrect decision is made.

B: The company is innocent and the jury acquits it. A correct decision is made.

C: The company is not innocent and the jury acquits it. An incorrect decision is made.

D: The company is not innocent and the jury finds it guilty. A correct decision is made.

▶ Of the two incorrect decisions, scenario A is considered to be the more serious because our society never wants to punish an innocent person or entity. We try to make the size of "reasonable doubt" larger than some might like.

▶ In the language of statistics, the initial hypothesis is called the null hypothesis, often designated as H_0. Competing with this is the alternative hypothesis, H_a. The amount of reasonable doubt is called the **level of significance**, represented by α. As we did with our study of confidence intervals, we are only going to consider scenarios that lend themselves to using the normal distribution. Reasonable doubt is any case in which the offered alternative can be explained as a reasonable deviation from a stated mean or proportion.

▶ Let's go back to the discussion of the amount of advertisements being played by the radio station and examine the three alternatives we discussed.

▶ **Case 1**

H_0: The average amount of ad time per hour is 20 minutes.

H_a: The average amount of ad time per hour is less than 20 minutes.

▶ The graph for the problem is shown below. The shaded region represents the area of reasonable doubt, while the unshaded region represents a region where natural variation from the mean is not reasonable. (Please note that if the ad time for this station is actually more than 20 minutes per hour, the null hypothesis is still accepted.) This is called a left tail test because the unshaded region is on the left side of the graph.

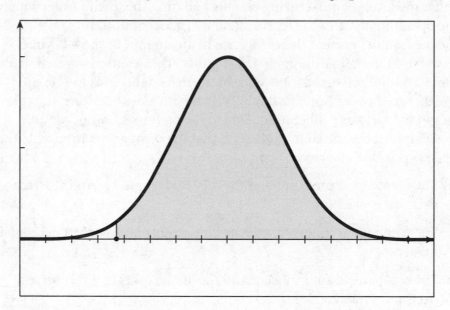

▶ **Case 2**

H_0: The average amount of ad time per hour is not 20 minutes.

H_a: The average amount of ad time per hour is greater than 20 minutes.

▶ The graph for the problem is shown below. The shaded region represents the area of reasonable doubt, while the unshaded region represents a region where natural variation from the mean is not reasonable. (Please note, that if the ad time for this station is actually less than 20 minutes per hour, the null hypothesis is still accepted.) This is a right tail test.

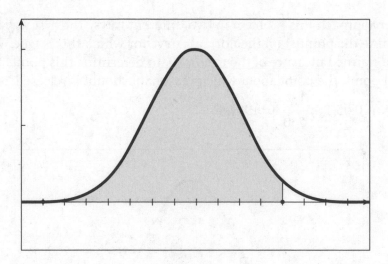

▶ **Case 3**

   H_0: The average amount of ad time per hour is 20 minutes.

   H_a: The average amount of ad time per hour is not 20 minutes.

▶ The graph for the problem is shown below. The shaded region represents the area of reasonable doubt, while the unshaded region represents a region where natural variation from the mean is not reasonable. Note that each unshaded tail contains one-half the value of α. This is a two tail test.

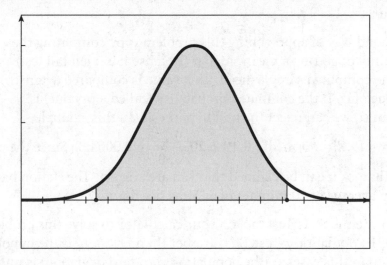

EXAMPLE

▶ Suppose that your friend collects data from 50 randomly selected one-hour intervals and determines that the mean number of minutes of ad time is 19.2 minutes with a standard deviation of 1.7 minutes. Let's put the level of significance at 5%. The area of the unshaded region is 0.05.

▶ We can approach this problem in two different ways. The first way is to determine the point(s) on the normal curve for which this is true. We use the claimed measure of the mean, 20, to determine this point. The critical point, the point about which reasonable doubt is released, is

$$invNorm\left(0.05, 20, \frac{1.7}{\sqrt{50}}\right) = 19.6046.$$

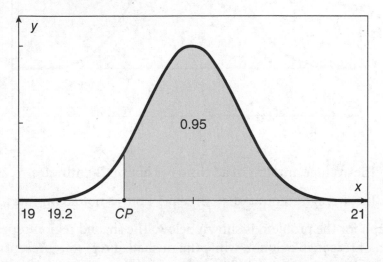

▶ Since the sample mean is less than this value, it can be claimed that you are correct. The station has less than an average of 20 minutes of ad time per hour.

▶ The second way of approaching this problem is by computing the probability of getting a mean smaller (because it is a left tail test) than the computed sample mean. This value is compared against the value of α. If the computed probability (called a p-value) is less than α, we can reject the null hypothesis. In this example,

$$p = P(\overline{x} < 19.2) = \text{normcdf}\left(0, 19.2, 20, \frac{1.7}{\sqrt{50}}\right) = 0.000438.$$ Since the p-value

is less than α, it can be claimed that you are correct. The station has less than an average of 20 minutes of ad time per hour.

▶ We can use the Stats Test tool on your calculator to solve this problem. Choose 1 Population z-Test, Stats (rather than Data), μ_0 is the hypothesized mean (20 in this case), σ (the population standard deviation) is unknown, so use the sample standard deviation 1.7, the sample mean is 19.2, n is 50, and use the drop menu for the Alternative Hypothesis to read $H_a : \mu < \mu_0$.

zTest 20,1.7,19.2,50,−1: *stat.results*

"Title"	"z Test"
"Alternate Hyp"	"$\mu < \mu 0$"
"z"	−3.32756
"PVal"	0.000438
"$\overline{x}$"	19.2
"n"	50.
"σ"	1.7

▶ You can see that the p-value is exactly what we calculated. The critical point is given as the z-score –3.32756. You can convert this back to raw data if you wish to compare the number of minutes involved. In either case, you are able to make a decision.

EXAMPLE

▶ Hardly's Beef and Shake claims that its half-pound burgers contain 45 g of fat. A random sample of 60 of its half-pound burgers contains an average of 46.1 g of fat with standard deviation 0.3 grams. Does this data support the claim that the half-pound burgers at Hardly's Beef and Shake have at most 45 g of fat? Use a 5% level of significance.

▶ The null hypothesis is that the mean is less than or equal to 45 $\left(H_0 : \mu \le 45\right)$ while the alternative hypothesis is that the mean is greater than 45 $\left(H_a : \mu > 45\right)$. Because the alternative hypothesis claims the mean to be a larger value than claimed by the manufacturer, we use a right tail test (the critical point will be at the right end of the normal distribution).

Critical Point Approach. The critical point for this test is

$$\text{invNorm}\left(0.95, 45, \frac{0.3}{\sqrt{60}}\right) = 45.0637.$$ Since the sample mean exceeds

the critical point, it can be claimed that the mean number of grams of fat in a half-pound burger from Hardly's Beef and Shake is greater than 45.

Calculator's Stat Test. Make sure you set the Alternative Hypothesis to $H_a : \mu > \mu_0$.

zTest 45,0.3,46.1,60,1: *stat.results*

"Title"	"z Test"
"Alternate Hyp"	"$\mu > \mu 0$"
"z"	28.4019
"PVal"	1.07733E-177
"$\bar{x}$"	46.1
"n"	60.
"σ"	0.3

The p-value is extremely small, well below the value of α. Consequently, reject H_0 and accept the claim that the fat content is greater than 45 g.

EXAMPLE

Conventional wisdom had been that 90% of the graduating class of Mendocino-Nueces County high schools went on to continue their education at a four-year college or university. A random sample of 250 graduating classes over the past few years shows that the proportion of graduates going to a four-year program is 89.2%. At the 5% level of significance, is this rate different from that of 90%?

The null hypothesis is that the proportion is 90% ($H_0: p = 0.90$), and the alternative is that it is not ($H_a: p \neq 0.90$). Without claiming to be higher or lower, this creates a two-tail test so that each tail on the bell curve will contain an area of 0.025.

Critical Point Approach. The left critical point is

$$\text{invNorm}\left(0.025, 0.9, \sqrt{\frac{(0.9)(0.1)}{250}}\right) = 0.862812 \text{ while the right-hand}$$

critical point is $\text{invNorm}\left(0.975, 0.9, \sqrt{\frac{(0.9)(0.1)}{250}}\right) = 0.937188.$ With

the sample proportion at 0.87, there does not appear to be enough evidence to claim that the proportion has changed.

With this being a two tail test, the area in each tail will be $\frac{\alpha}{2} = 0.025$. The results from the Stat Tool are:

zTest_1Prop 0.9,223,250,0: *stat.results*

$$\begin{bmatrix} \text{"Title"} & \text{"1}-\text{Prop z Test"} \\ \text{"Alternate Hyp"} & \text{"prop} \neq \text{p0"} \\ \text{"z"} & -0.421637 \\ \text{"PVal"} & 0.67329 \\ \text{"}\hat{p}\text{"} & 0.892 \\ \text{"n"} & 250. \end{bmatrix}$$

The p-value (0.67329) is larger than the value of alpha (0.05) indicating that the sample statistic is not significantly different than the hypothesized value of 0.9. We fail to reject the null hypothesis. (We never say that we accept the null hypothesis in the same way a jury never finds the defendant innocent. All that has happened is that not enough evidence has been produced to reach the point "beyond a reasonable doubt.")

EXAMPLE

The chancellor of a large city's school system claims that at least 65% of all the juniors in the city's high schools are enrolled in an Algebra II course. A random sample of 80 of the city's high schools shows that 52% of the juniors are enrolled in an Algebra II course. Does this result contradict the statement made by the Chancellor? Test the claim at the 2% level of significance.

The null hypothesis is that the proportion is greater than or equal to 0.65, while the alternative hypothesis is that the proportion is less than 0.65. That is, $H_0: p \geq 0.65$ and $H_a: p < 0.65$. Use a left tail test.

Critical Point Approach. The critical point is

$$\text{invNorm}\left(0.02, 0.65, \sqrt{\frac{(0.65)(0.35)}{80}}\right) = 0.54048.$$ With the sample

proportion exceeding this critical value, there is not enough evidence to challenge the Chancellor's claim.

P-Value Approach. The results of the Stat Tool on your calculator are:

zTest_1Prop 0.65,42,80,0: *stat.results*

"Title"	"1−Prop z Test"
"Alternate Hyp"	"prop ≠ p0"
"z"	−2.34404
"PVal"	0.019076
"p̂"	0.525
"n"	80.

The p-value (0.675962) is greater than the value of α, so we fail to reject the null hypothesis. Not enough data has been shown to change the claim that at least 65% of the juniors are taking Algebra II.

EXERCISE 11.5

For each problem, state the null and alternative hypotheses, the value of the critical point and the p-value, the decision that should be and an interpretation of the result.

1. A manufacturer claims that the mean weight of his five-pound bag of flour is 5.02 pounds. A random sample of 50 bags of flour gives a mean weight of 4.99 pounds with a standard deviation of 0.03 pounds. Does the data provide significant results to refute the manufacturer's claim? Use the 5% level of significance.

2. An old commercial claimed, "Three out of five dentists recommend that you choose sugarless gum." Wondering if this is still a valid statement, a survey was taken of 4,000 randomly selected dentists around the country. The results show that 2,225 of the dentists recommend sugarless gum. Does the result of the survey indicate that there has been a change to the recommendation for using sugarless gum? Use the 3% level of significance.

3. Management claims that the average number of hours of overtime given each week to workers has decreased in the past year. The mean number of overtime hours per week for the last three years has been 30.7 with a standard deviation of 2.8 hours. A random sample of 25 weeks from the last year shows a mean of 31.1 hours of overtime. Does the data support management's claim at the 2% level of significance?

4. Mike and Jack were listening to the radio while they were working. Mike claimed that at least half the songs in rock and roll were about love. Jack disagreed with this statement and claimed he thought it was less than that. After they decided how to define a song involving love, they kept a tally of the songs heard from the various radio stations and by Internet services they were using—being careful not to include the same song twice in their survey. Of the 75 songs they heard, 36 were classified as being about love. Does this data support Jack's claim? Use the 1% level of significance.

SIMULATION

Simulation is a technique that uses a probabilistic model to test a theory. For example, a supposedly "fair" coin showed heads 5 times in a row when flipped. Does 5 times in a row give an indication that the coin might be biased? Rather than take the time to collect a large number of samples of 5 flips of a coin, it is possible to create a computer program to simulate these flips using a random number generator and have the program record the number of heads. Use the graphing calculator to generate a set of random numbers. (Be sure to seed the random number generator. Read the manual for your calculator to learn how this is done.) Designate 0 to represent tails and 1 to represent heads. The command randint(0,1,5) will generate a set of 5 numbers, 0 or 1. Count the number of times one appears to represent the number of heads on that trial. A program was used to represent 2,000 repetitions of this action.

The accompanying figure illustrates the outcome of 2,000 flips of 5 fair coins. Fifty-eight out of the 2,000 outcomes contain 5 heads. We know the probability of getting 5 heads in 5 flips is fairly small, $\binom{5}{5}\left(\frac{1}{2}\right)^5 = \frac{1}{32}$, it is not "impossible." The result of the simulation is that we do not have sufficient evidence to doubt the coin is fair.

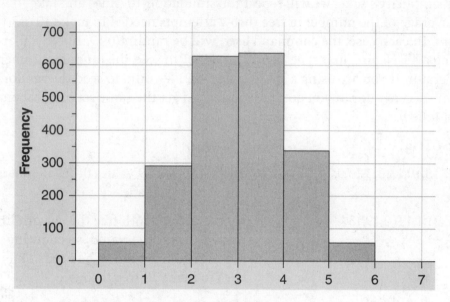

What happens if instead of a streak of 5 heads in a row, there is a streak of 15 heads. Should this give you pause to believe that the coin might be biased? Executing the simulation for flipping 15 coins 2,500 times we find there are 3 cases in which all 15 outcomes are heads.

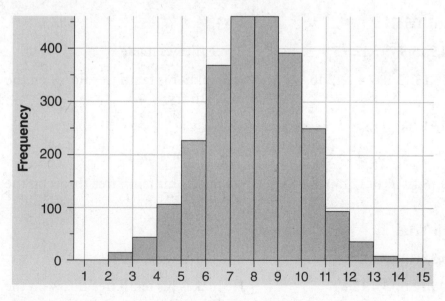

The probability of this happening is $\binom{15}{15}\left(\frac{1}{2}\right)^{15} = \frac{1}{32,768}$, a very small number but still not equal to 0.

Colin plays recreational basketball, and the percent of free throws made is 85% this season. Create a simulation to estimate the number of free throws he will need to shoot in order to make 10 more free throws.

With his average being 85%, we can designate the digits 0–16 as indicating he made a free throw and 17–19 that he did not. We can generate sets of random numbers using our graphing calculator to get our results. For simplicity's sake, we will repeat the simulation 10 times and take an average of the number of free throw attempts needed to make 10 of them. In each case, the command used will be randint(0,19, 10). If you are using a TI-Nspire, first type RandSeed 54321 to seed the random number generator. If you are using a TI-84 type 54321 → rand to seed the random number generator. If you do this, you should get the same set of numbers that follow.

First Trial

12,7,15,19,13,3,2,13,9,2

There a total of 9 values 0–16, so that is 9 free throws made.

5,8,6,1,1,1,9,0,19,13

He makes his tenth free throw with his first throw in this round, so counting the 10 tries from line 1, he needs 11 tries to make 10 free throws.

Second Trial

12,14,11,14,5,9,4,10,11,17

9 free throws made

15,11,6,2,10,6,2,4,17,4

He makes his tenth free throw on the eleventh try.

Third Trial

16,4,9,15,8,7,13,17,6,3

9 free throws made

14,13,13,13,18,7,4,2,12,10

He makes his tenth free throw on the eleventh try.

Fourth Trial

6,6,19,11,5,5,3,6,17,5

8 free throws made

16,**9**,16,19,16,10,17,15,11,19

He makes his tenth free throw on the twelfth try

Fifth Trial

17,0,3,4,14,6,17,12,1,15

8 free throws made

7,**0**,19,12,6,15,4,12,8,8

He makes his tenth free throw on the twelfth try

Sixth Trial

14,1,13,7,16,4,9,6,11,**15**

10 tries needed to make 10 free throws

▸ **Seventh Trial**

▸ 19,8,17,5,10,4,9,12,0,17   7 free throws made

▸ 19,17,14,19,19,3,**8**,5,7,12   He makes his tenth free throw on the
                                          seventeenth try

▸ **Eighth Trial**

▸ 10,0,7,0,13,7,12,18,13,4   9 free throws made

▸ **11**,11,8,9,9,13,14,11,4,13   He makes his tenth free throw on the
                                          eleventh try

▸ **Ninth Trial**

▸ 11,16,16,5,16,6,0,5,9,**11**   10 tries needed to make 10 free throws

▸ **Tenth Trial**

▸ 17,10,3,3,13,17,17,4,14,   7 free throws made

▸ 2,8,19,**2**,15,0,14,9,3,7   He makes his tenth free throw on the
                                          fourteenth try

▸ According to this simulation, it will take Colin an average of 11.9 free throws in order to make 10 of them.

▸ Of course, to get a better estimate, use a computer to repeat this process a large number of times.

▸ The simulation could also be done without a computer. You could use a jar with a large number of marbles. For example, have a jar with 85 blue and 15 red marbles. Randomly select a marble from the jar. Continue to do so until 10 blue marbles are drawn. Record the total number of marbles drawn from the jar to represent the number of free throws attempted. Return the marbles to the jar. Shake the jar to redistribute the marbles in the jar before collecting the data for the next trial.

EXAMPLE

▸ Carson has a ham and cheese sandwich for school lunch 35% of the time, and he does so on a random basis. Describe a simulation that can be used to determine the number of times Carson has a ham and cheese sandwich for lunch during a month in which there are 20 school days.

▸ Designate 0–34 as days that Carson has a ham and cheese sandwich for lunch and 35–99 as days that he does not. Seed the random number generator to ensure randomness. A command such as randint(0,99,20) can represent 1 trial. Tally the number of times the numbers 0–34 appear in each trial.

EXERCISE 11.6

Describe a simulation that can be used to determine an answer to each of these problems.

1. The makers of Ty Rex cereal advertise that you get a plastic dinosaur with every box of cereal that you buy. There are eight different dinosaurs in all. How many boxes of cereal would you expect to buy, on average, to get the complete set of dinosaurs?

2. The teams in the NBA Western Conference are considered to be better than the teams in the Eastern Conference. In a given year, the teams from the West beat the East in 60% of the games played. The final round of the NBA playoffs has a team from each division play each other in a best 4 out of 7 series. Determine the average number of games that need to be played to determine a winner.

An Introduction to Matrices

It is not uncommon for a car dealership to have multiple stores across a geographic region. Consider the case of the Treanor Brothers automobile dealership with stores in Charlotte, NC; Charleston, SC; Myrtle Beach, SC; and Concord, NC. The November inventory is taken of five of its bestselling SUVs—Toyota Highlander, Lexus RX, Buick Enclave, Ford Explorer, and the BMW X5. The table below shows the inventory for each model at each location.

	Toyota Highlander	Lexus RX	Buick Enclave	Ford Explorer	BMW X5
Charlotte, NC	12	21	15	19	14
Charleston, SC	10	18	12	15	21
Myrtle Beach, SC	7	5	21	23	28
Concord, NC	9	11	23	17	13

The wholesale price of each model, the price the Treanor Brothers paid for each model, is given in the next table.

Model	Wholesale Price
Toyota Highlander	$38,141
Lexus RX	$42,404
Buick Enclave	$39,195
Ford Explorer	$31,350
BMW X5	$56,056

▶ Compute the value of the SUV inventory for the month of November at each of the Treanor Brothers' four stores.

▶ For each store, multiply the number of each model by the wholesale cost of the model:

Charlotte: 12(38,141) + 21(42,404) + 15(39,195) + 19(31,350)+ 14(56,056) = $3,316,535

Charleston: 10(38,141) + 18(42,404) + 12(39,195) + 15(31,350)+ 21(56,056) = 3,262,448

Myrtle Beach: 7(38,141) + 5(42,404) + 21(39,195) + 23(31,350)+ 28(56,056) = 3,592,720

Concord: 9(38,141) + 11(42,404) + 28(39,195) + 17(31,350)+ 13(56,056) = 2,972,876

We will now use this example to illustrate the mathematical construct called a **matrix**. A matrix is a rectangular array of numbers. Ignoring the labels, which are included to help read the tables, the first table has four rows and five columns. Rows are read horizontally and columns vertically. The second table has 5 rows and 1 column. The number of rows and the number of columns identifies the **dimensions of a matrix**. Let **A** represent the inventory and **B** the wholesale value. The dimensions of **A** are 4x5, and the dimensions of **B** are 5x1.

$$\mathbf{A} = \begin{bmatrix} 12 & 21 & 15 & 19 & 14 \\ 19 & 18 & 12 & 15 & 21 \\ 7 & 5 & 21 & 23 & 28 \\ 9 & 11 & 23 & 17 & 13 \end{bmatrix} \text{ and } \mathbf{B} = \begin{bmatrix} 38,141 \\ 42,404 \\ 39,195 \\ 31,350 \\ 56,056 \end{bmatrix}$$

Multiply matrices **A** and **B** to get:

$$\mathbf{AB} = \begin{bmatrix} 3,316,535 \\ 3,262,448 \\ 3,592,720 \\ 2,972,876 \end{bmatrix}$$

These are the same values calculated in the solution to the first example, the total value of the inventory for each store. This tells us the process in which multiplication of matrices occurs. Starting with row 1 of the left-hand factor, multiply each element in row 1 with its corresponding element from column 1 in the right-hand factor, and add these products to get the result. It is important to note that a requirement for matrix multiplication is that the number of columns in the left-hand factor must equal the number of rows in the right-hand factor.

Matrix multiplication and the inverse of a matrix can be used to solve systems of linear equations. For instance, consider the system:

$$2x - 3y = 37$$

$$5x + 4y = 12$$

This is the equivalent of the matrix equation:

$$\begin{bmatrix} 2 & -3 \\ 5 & 4 \end{bmatrix}\begin{bmatrix} x \\ y \end{bmatrix} = \begin{bmatrix} 37 \\ 12 \end{bmatrix}$$

Observe that the first matrix contains the coefficients from the system. The second matrix contains the variables of the system. The third system contains the constants of the linear equations. When you perform the matrix multiplication on the left, you see the same terms as the left side of the system of equations.

$$\begin{bmatrix} 2x - 3y \\ 5x + 4y \end{bmatrix} = \begin{bmatrix} 37 \\ 12 \end{bmatrix}$$

This matrix equation can be written as [A] [X] = [B]. In Algebra 1, we would solve the equation $ax = b$ by dividing both sides of the equation by a to get $x = \dfrac{b}{a}$. However, we can't divide matrices. We could also solve the equation by multiplying both sides of the equation by the multiplicative inverse of a, $\left(\dfrac{1}{a}\right)(ax) = \left(\dfrac{1}{a}\right)b$, which gives the same result, $x = \dfrac{b}{a}$. If you choose to, you can go online to read that multiplication of matrices is *not* a commutative operation, so the order of operation must be done carefully. For the purposes of solving the system of equations, you need to multiply the inverse as the left factor. (If you try to put the inverse as a right factor, the calculator will give you an error message.)

[A] [X] = [B]

[A]⁻¹ [A] [X] = [A]⁻¹ [B]

[X] = [A]⁻¹ [B]

The solution to the system of equations is the point (8, –7).

$$\begin{bmatrix} x \\ y \end{bmatrix} = \begin{bmatrix} 2 & -3 \\ 5 & 4 \end{bmatrix}^{-1}\begin{bmatrix} 37 \\ 12 \end{bmatrix}$$

$$\begin{bmatrix} x \\ y \end{bmatrix} = \begin{bmatrix} 8 \\ -7 \end{bmatrix}$$

EXERCISE A.1

Solve each system of equations using matrix algebra.

1. $6x + 4y = 0$
   $4x + 9y = 95$

2. $2x + 3y - 2z = 15$
   $3x - 4y + 2z = 16$
   $7x + 10y + 5z = 216$

Conditional and Binomial Probabilities

It will be worthwhile to review some of the concepts of probability before we look at descriptive and inferential statistics. As you recall, the probability of an event is the relative frequency with which the event will occur. The probability of rolling a 3 from a fair die is $\frac{1}{6}$ because only one side out of the six shows a 3. The probability of randomly drawing an ace from a well-shuffled deck of bridge cards is $\frac{4}{52} = \frac{1}{13}$ because 4 of the 52 cards in the deck are aces.

Conditional Probability

Conditional probability asks for the probability of an event happening after some other event has already occurred. Let's examine the question about the ace a little bit. What is the probability of getting an ace from a well-shuffled bridge deck on the second pick if the first card picked is not an ace and is not returned to the deck? We are told that a card from the deck was selected and that card was not an ace. This tells us that the deck now has

51 cards and 4 of those are aces. Therefore, P(second card is an ace| first card is not an ace) = $\frac{4}{51}$. (This reads, the probability that the second card is an ace given that the first card is not an ace is $\frac{4}{51}$.)

EXAMPLE

▶ A fair die (6 sides—all with the same chance of facing up) is rolled twice. What is the probability that the second roll is a 6 given that the first roll is a 5?

▶ Unlike the pulling of a card from a deck and not returning the card, the chance of rolling a 6 is the same whether it is the first roll or the tenth roll.

$$P(6|5) = \frac{1}{6}.$$

Two events, A and B, are said to be **independent** of each other if $P(B|A) = P(B)$. That is, the first outcome has no impact on the opportunity for the second outcome to occur.

EXAMPLE

▶ The table below shows the result of a random sample of 200 voters. The voters were asked to identify how they were registered with the local Board of Elections.

	Female	Male	Total
Democrat	40	40	80
Republican	21	39	60
Independent	39	21	60
Total	100	100	200

(a) If a voter from this survey is selected at random, what is the probability that the voter is a registered Democrat?

(b) If a voter from this survey is selected at random, what is the probability that the voter is a female?

(c) If a voter from this survey is selected at random, what is the probability that the voter is a registered Democrat given that the voter is a woman?

(d) If a voter from this survey is selected at random, what is the probability that the voter is a woman given that the voter is a registered Democrat?

Solutions:

(a) 80 of the people in the survey are registered Democrats, so
$$P(\text{Democrat}) = \frac{80}{200}.$$

(b) 100 of the people survey are women so $P(\text{Female}) = \frac{100}{200}.$

(c) There are 100 women in the survey, of whom, 40 are Democrats.
$$P(\text{Democrat} \mid \text{Female}) = \frac{40}{100}.$$

(d) Of the 80 Democrats, 40 are women. $P(\text{Female} \mid \text{Democrat}) = \frac{40}{80}.$

The two events "the voter is a woman" and "the voter is a registered Democrat" are independent of each other because $P(\text{Female}) = P(\text{Female} \mid \text{Democrat})$.

A second test for independence is the rule: if $P(A \text{ and } B) = P(A) \times P(B)$, then A and B are independent.

EXAMPLE

Example: The results of a survey of the junior class at Central High School are shown in the Venn diagram.

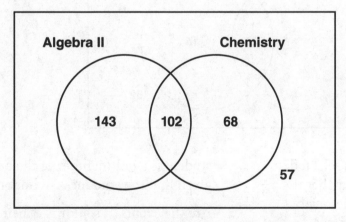

(a) How many students are in the junior class of Central High School?

(b) If a junior from Central High School is selected at random, what is the probability the student is enrolled in an Algebra II class?

(c) If a junior from Central High school is selected at random, what is the probability the student is enrolled in a Chemistry class?

(d) Are the events "the student is enrolled in Algebra II" and "the student is enrolled in Chemistry" independent events for the juniors at Central High School?

Solutions:

(a) There are 143 + 102 + 68 + 57 = 370 juniors in Central High School.

(b) $P(\text{Algebra II}) = \dfrac{143+102}{370} = \dfrac{245}{370} = 0.662.$

(c) $P(\text{Chemistry}) = \dfrac{170}{370} = 0.459$

(d) $P(\text{Algebra II and Chemistry}) = \dfrac{102}{370} = 0.276.$ $P(\text{Algebra II}) \times$

$P(\text{Chemistry}) = \left(\dfrac{170}{370}\right)\left(\dfrac{102}{370}\right) = \dfrac{41650}{136900} = 0.304.$ Since these two probabilities are not equal, the events are not independent.

EXERCISE B.1

Use the accompanying Venn diagram, which represents the members of a senior class at a small high school, to answer questions 1–3.

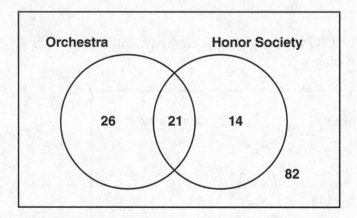

1. If a member of the senior class is selected at random, what is the probability that the student is a member of the honor society given that the student is in orchestra?

2. If a member of the senior class is selected at random, what is the probability that the student is a not member of the orchestra given that the student is in honor society?

3. Are the events "a senior is a member of the honor society" and "a senior is a member of the orchestra" independent of each other?

Use the accompanying table that represents the amount of time a person listens to music each day to answer questions 4–6.

Age	1–2 hrs	2–3 hrs	4+ hrs	Total
15–17	20	23	29	72
18–20	29	22	21	72
21–24	27	34	17	78
Total	76	79	67	222

4. What is the probability that a randomly selected person from this survey listens to music 1–2 hours each day if that person is between the ages of 15 and 17?

5. What is the probability that a randomly selected person from this survey is between the ages of 18 and 20 if that person listens to music for 4 or more hours each day?

6. Are the events "the person is between 21 and 24 years old" and "the person listens to music 2–3 hours each day" independent of each other?

Bernoulli Trials/Binomial Probability

A probability experiment with the properties: there are n independent trials; there are two outcomes per trial (success and failure); and P(success) is constant from trial to trial is called a binomial experiment. (It is also called a Bernoulli trial after the family of Swiss mathematicians.) The probability

of exactly r successes in n trials is given by the formula $\begin{pmatrix} n \\ r \end{pmatrix} p^r (1-p)^{n-r}$,

where p represents the probability of success on any given trial.

EXAMPLE

A fair 6-sided die is rolled 5 times. What is the probability of getting a 2 four times?

If success is defined as "the die shows a 2" and failure by "the die does not show a 2," then this is a binomial experiment. (This is key: it is not important if the roll is a 1 or a 5—the only thing that matters is that it is not a 2.)

$$P(r=4) = \begin{pmatrix} 5 \\ 4 \end{pmatrix} \left(\frac{1}{6}\right)^4 \left(\frac{5}{6}\right)^1 = \frac{25}{7776} = 0.003125$$

There are functions on most graphing calculators called binomPdf that will compute the probability of a specific number of events. $P(r = 4) =$

$\text{binomPdf}\left(5, \frac{1}{6}, 4\right) = 0.003125.$

EXAMPLE

It is found that 2% of all grommets produced by the Acme Corporation are defective. In a random sample of 1,000 grommets produced by Acme, what is the probability that at most 3 of them are defective?

This is also a binomial experiment because the chance of a grommet being defective is independent of whether the previous grommet tested was defective. Success in this experiment is defined by finding a defective grommet. The probability that at most 3 grommets are found to be defective is $P(r \leq 3)$ and this is equal to $P(r = 0) + P(r = 1) + P(r = 2) + P(r = 3)$. With $n = 1000$ and

$$p = .02, = \binom{1000}{0}(0.02)^0(0.98)^{1000} + \binom{1000}{1}(0.02)^1(0.98)^{999} +$$

$$\binom{1000}{2}(0.02)^2(0.98)^{998} + \binom{1000}{3}(0.02)^3(0.98)^{997}.$$ There are functions

on most graphing calculators called binomCdf that will compute the probability of a range of events. For this problem, the solution is binomCdf(1000,0.02,0,3) = 0.000003.

EXAMPLE

Example: The spinner shown in the accompanying diagram is spun 10 times. What is the probability of getting the outcome Green at least 3 times?

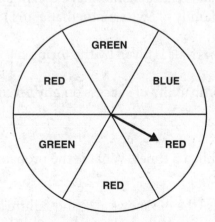

The board does not change shape, so P(Green) is consistently $\frac{1}{3}$. $P(r \geq 3)$ = $P(r = 3) + P(r = 4) + P(r = 5) + \ldots + P(r = 10)$. This is a lot of computing to do. The complement of this event—the outcomes that are of interest— are $P(r \leq 2) = P(r = 0) + P(r = 1) + P(r = 2)$. Take advantage of the fact that P(event) + P(event's complement) must equal 1 because, together, the event and its complement make up all the possible outcomes for an experiment. $P(r \geq 3) = 1 - P(r \leq 2) = 1 - \text{binomCdf}\left(10, \frac{1}{3}, 0, 2\right) = 0.9827$, answer rounded to four decimal places.

EXERCISE B.2

Use the diagram of the spinner in the section to answer questions 1–3.

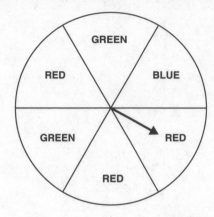

1. If the spinner is spun 10 times, what is the probability of getting the result red three times?

2. If the spinner is spun 5 times, what is the probability of getting the result blue at most twice?

3. If the spinner is spun 8 times, what is the probability of getting the result green at least twice?

4. Based on past statistics, a company knows that 99% of its ball bearings pass a quality control test. A random sample of 100 ball bearings is tested for the quality control test. What is the probability that all 100 bearings will pass the test?

One percent of the widgets produced by the GernX Manufacturing Company fail to comply with company standards. A random sample of 50 widgets is selected from a production run.

5. What is the probability that at most 3 of the widgets will fail to meet company standards?

6. What is the probability at least 99 of the widgets will fail to meet company standards?

Answer Key

CHAPTER 1

Linear Equations and Inequalities

EXERCISE 1.1

1. 9.8

$$3x-19=8x-68$$
$$49=5x$$
$$x=9.8$$

2. 4.92

$$53-14x+21=11x-49$$
$$74-14x=11x-49$$
$$123=25x$$
$$x=4.92$$

3. 44

$$(13)(17)(4)\left(\frac{x+21}{13}+\frac{2x-3}{17}\right)=(13)(17)(4)\left(\frac{x-4}{4}\right)$$
$$68(x+21)+52(2x-3)=221(x-4)$$
$$68x+1,428+104x-156=221x-884$$
$$172x+1,272=221x-884$$
$$2,156=49x$$
$$x=44$$

4. $\dfrac{31b}{11}$

$$60a\left(\frac{x+b}{3a}+\frac{x-b}{4a}\right)=60a\left(\frac{2x+3b}{5a}\right)$$
$$20(x+b)+15(x-b)=12(2x+3b)$$
$$20x+20b+15x-15b=24x+36b$$
$$35x+5b=24x+36b$$
$$11x=31b$$
$$x=\frac{31b}{11}$$

5. 80 @ \$4 and 40 @ \$6

p: number of \$4 plants purchased

$120 - p$: number of \$6 plants purchased

$$4p+6(120-p)=560$$
$$4p+720-6p=560$$
$$-2p=-160$$
$$p=80$$

6. Pre 1970: 40; 1970 to 2000: 140; after 2000: 67

n: number of movies from pre-1970

$3n + 20$: number of movies from 1970 to 2000

$n + 27$: number of movies from after 2000

$n + 3n + 20 + n + 27 = 247$

$5n + 47 = 247$

$5n = 200$

$n = 40$

EXERCISE 1.2

1. $x < 11; (-\infty, 11)$

$22 > 2x$

$11 > x$

2. $x \geq 14; [14, \infty)$

$6x - 14 - 15 + 12x \geq 15x + 13$

$18x - 29 \geq 15x + 13$

$3x \geq 42$

$x \geq 14$

3. $9 \leq x < 16; [9, 16)$

$27 \leq 3x < 48$

$9 \leq x < 16$

4. $\frac{2}{5} \leq x < 4; \left[\frac{2}{5}, 4\right)$

$-20 < -5x \leq -2$

$4 > x \geq \frac{2}{5}$

5. $x < 3$ or $x \geq 5; (-\infty, 3) \cup [5, \infty)$

$3x < 9$ or $5x \geq 25$

$x < 3$ or $x \geq 5$

6. Real Numbers $(x > -3$ or $x < 5)$

$-5x < 15$ or $3x < 15$

$x > -3$ or $x < 5$

7. 25 ounces

c: number of ounces of cashews added to the mixture

$$\frac{\text{cashews}}{\text{total weight}} : \frac{20 + c}{50 + c} \geq \frac{3}{5}$$

$5(20 + c) \geq 3(50 + c)$

$100 + 5c \geq 150 + 3c$

$2c \geq 50$

$c \geq 25$

EXERCISE 1.3

1. $(3, -7)$

Substitute $-3x + 2$ for y

$-3x + 2 = 2x - 13$

$-5x = -15$

$x = 3$

$y = -3(3) + 2 = -7$

2. $(-5, 2)$

Substitute $x + 7$ for y

$x + 7 = 4x + 22$

$-15 = 3x$

$x = -5$

$y = -5 + 7 = 2$

3. $(-10, 12)$

Substitute $x + 22$ for y

$2x + 3(x + 22) = 16$

$2x + 3x + 66 = 16$

$5x = -50$

$x = -10$

$y = -10 + 22 = 12$

4. $(6.3, 5.2)$

Substitute $x - 1.1$ for y

$5x + 4(x - 1.1) = 52.3$

$5x + 4x - 4.4 = 52.3$

$9x = 56.7$

$x = 6.3$

$y = 6.3 - 1.1 = 5.2$

5. $(-12.2, 13.8)$

Substitute $x + 26$ for y

$7x - 2(x + 26) = -113$

$7x - 2x - 52 = -113$

$5x = -61$

$x = -12.2$

$y = -12.2 + 26 = 13.8$

6. $(7, -3)$

$3(3x - 5y = 36)$

$5(2x + 3y = 5)$

becomes:

$9x - 15y = 108$

$10x + 15y = 25$

Add:

$19x = 133$

$x = 7$

$2(7) + 3y = 5$

$3y = -9$

$y = -3$

7. $(-8, 5)$

$2(-4x + 7y = 67)$

$7(-3x - 2y = 14)$

becomes:

$-8x + 14y = 134$

$-21x - 14y = 98$

Add:

$-29x = 232$

$x = -8$

$-4(-8) + 7y = 67$

$32 + 7y = 67$

$7y = 35$

$y = 5$

8. $(-15, -45)$

$-3(5x - 2y = 15)$

$2(4x - 3y = 75)$

becomes:

$-15x + 6y = -45$

$8x - 6y = 150$

Add:

$-7x = 105$

$x = -15$

$5(-15) - 2y = 15$

$-75 - 2y = 15$

$-2y = 90$

$y = -45$

9. $\left(\dfrac{-5}{6}, \dfrac{2}{3}\right)$

$2(18x + 9y = -9)$

$-3(12x + 24y = 6)$

becomes:

$36x + 18y = -18$

$-36x - 72y = -18$

Add:

$-54y = -36$

$y = \dfrac{2}{3}$

$18x + 9\left(\dfrac{2}{3}\right) = -9$

$18x + 6 = -9$

$18x = -15$

$x = \dfrac{-5}{6}$

10. No solution

$-4(3x + 4y = 42)$

$\quad 12x + 16y = 21$

becomes:

$-12x - 16y = -168$

$\quad 12x + 16y = 21$

Add:

$0 = -147$ which is a false statement

11. Infinite set of solutions

$-3(5x - 3y = 25)$

$\quad 15x - 9y = 75$

becomes:

$-15x + 9y = -75$

$\quad 15x - 9y = 75$

Add:

$0 = 0$ which is a true statement

12. 20 pounds cashews and 30 pounds almonds

a: number of pounds of almonds added

c: number of pounds of cashews added

$\dfrac{a}{a+c} = \dfrac{3}{5}$ and $\dfrac{a+10}{a+10+c+20} = \dfrac{1}{2}$

Becomes:

$5a = 3(a+c)$ and $2(a+10) = a+c+30$

Simplify:

$2a - 3c = 0$ and $a = c + 10$

Substitute:

$2(c+10) - 3c = 0$

$2c + 20 - 3c = 0$

$c = 20$

$a = 20 + 10 = 30$

EXERCISE 1.4

1. (3, 5)

Graph $y = 2x - 1$ and $y = \dfrac{3}{2}x + \dfrac{1}{2}$

2. (–4, –11)

Graph $y = x - 7$ and $y = \dfrac{2}{5}x - \dfrac{47}{5} = .4x - 9.4$

3. (2, 13)

Graph $y = \dfrac{5}{2}x + 8$ and $y = 3x + 7$

4. (12,5)

Graph $y = \dfrac{-3}{4}x + 14$ and $y = \dfrac{4}{3}x - 11$

5. Infinite number of solutions

Graph $y = -.4x + .1$ twice

EXERCISE 1.5

1. $(-5, 7, 10)$

Eliminate the z:

$4(4x - 5y + 3z = -25)$ $5(4x - 5y + 3z = -25)$

$3(7x + 8y - 4z = -19)$ $-3(-3x - 4y + 5z = 37)$

$16x - 20y + 12z = -100$ $20x - 25y + 15z = -125$

$21x + 24y - 12z = -57$ $9x + 12y - 15z = -111$

Add:

$37x + 4y = -157$ $29x - 13y = -236$

Eliminate the y:

$13(37x + 4y = -157)$

$4(29x - 13y = -236)$

Distribute:

$481x + 52y = -2,041$

$116x - 52y = -944$

Add:

$597x = -2,985$

$x = -5$

Find y:

$37(-5) + 4y = -157$

$-185 + 4y = -157$

$4y = 28$

$y = 7$

Find z:

$4(-5) - 5(7) + 3z = -25$

$-55 + 3z = -25$

$3z = 30$

$z = 10$

2. $(12, 15, 21)$

Eliminate the z

$2(8x - 7y + 3z = 54)$ $3(8x - 7y + 3z = 54)$

$-3(5x + 4y + 2z = 162)$ $10x + 12y - 9z = 111$

Distribute:

$16x - 14y + 6z = 108$ $24x - 21y + 9z = 162$

$-15x - 12y - 6z = -486$ $10x + 12y - 9z = 111$

Add:

$x - 26y = -378$ $34x - 9y = 273$

Eliminate the x:

$-34(x - 26y = -378)$

$34x - 9y = 273$

Distribute:

$-34x + 884y = 12,852$

$34x - 9y = 273$

Add:

$875y = 13,125$

$y = 15$

Find x:

$x - 26(15) = -378$

$x - 390 = -378$

$x = 12$

Find z:

$8(12) - 7(15) + 3z = 54$

$96 - 105 + 3z = 54$

$3z = 63$

$z = 21$

3. $(-4, -9, -11)$

Eliminate the z:

$7x - 5y - 2z = 39$ ⎫ $-6(7x - 5y - 2z = 39)$
$4x + 3y + 2z = -65$ $2x + 5y - 12z = 79$

Add:

$11x - 2y = -26$ $-40x + 35y = -155$

Eliminate the y:

$35(11x - 2y = -26)$

$2(-40x + 35y = -155)$

$385x - 70y = -910$
$-80x + 70y = -310$

Add:

$305x = -1,220$

$x = -4$

Find y:

$11(-4) - 2y = -26$

$-2y = 18$

$y = -9$

Find z:

$7(-4) - 5(-9) - 2z = 39$

$17 - 2z = 39$

$-2z = 22$

$z = -11$

4. $\left(\dfrac{-2}{3}, \dfrac{3}{4}, \dfrac{5}{6}\right)$

Eliminate z:

$9x + 8y - 12z = -10$ $9x + 8y - 12z = -10$
$3(5x + 12y + 4z = 9)$ $-1(9x - 20y - 12z = -31)$

Simplify:

$9x + 8y - 12z = -10$ $9x + 8y - 12z = -10$
$15x + 36y + 12z = 27$ $-9x + 20y + 12z = 31$

Add:

$24x + 44y = 17$ $28y = 21$

$$y = \frac{3}{4}$$

Find x:

$24x + 44\left(\dfrac{3}{4}\right) = 17$

$24x + 33 = 17$

$24x = -16$

$x = \dfrac{-2}{3}$

Find z:

$9\left(\dfrac{-2}{3}\right) + 8\left(\dfrac{3}{4}\right) - 12z = -10$

$-6 + 6 - 12z = -10$

$z = \dfrac{5}{6}$

5. Students in Advance \$4; Students at Door \$5; Adults \$7

t: price of a student ticket sold in advance

d: price of a student ticket sold at the door

a: price of an adult ticket

Friday : $250t+50d+300a=3,350$

Saturday : $600t+50d+500a=6,150$

Sunday : $50t+70d+250a=2,300$

Eliminate d:

$-1(250t+50d+300a=3,350)$ $\qquad$ $7(250t+50d+300a=3,350)$

$600t+50d+500a=6,150$ $\qquad$ $-5(50t+70d+250a=2,300)$

Simplify:

$-250t-50d-300a=-3,350$ $\qquad$ $1,750t+350d+2,100a=23,450$

$600t+50d+500a=6,150$ $\qquad$ $-250t-350d-1,250a=-11,500$

Add:

$350t+200a=2,800$ $\qquad$ $1,500t+850a=11,950$

Eliminate a:

$-85(350t+200a=2,800)$

$20(1,500t+850a=11,950)$

Simplify:

$-29,750t-17,000a=-238,000$

$30,000t+17,000a=239,000$

Add:

$250t=1,000$

$t=4$

Find a:

$350(4)+200a=2,800$

$1,400+200a=2,800$

$200a=1,400$

$a=7$

Find d:

$50(4)+70d+250(7)=2,300$

$200+70d+1,750=2,300$

$70d=350$

$d=5$

EXERCISE 1.6

1. (17, 29)

$$\begin{bmatrix} 6 & -7 \\ 5 & 3 \end{bmatrix}\begin{bmatrix} x \\ y \end{bmatrix}=\begin{bmatrix} -101 \\ -172 \end{bmatrix}$$

$$\begin{bmatrix} x \\ y \end{bmatrix}=\begin{bmatrix} 6 & -7 \\ 5 & 3 \end{bmatrix}^{-1}\begin{bmatrix} -101 \\ -172 \end{bmatrix}=\begin{bmatrix} 17 \\ 29 \end{bmatrix}$$

2. (35, 57)

$$\begin{bmatrix} 6 & -3 \\ 4 & 7 \end{bmatrix}\begin{bmatrix} x \\ y \end{bmatrix}=\begin{bmatrix} 39 \\ 539 \end{bmatrix}$$

$$\begin{bmatrix} x \\ y \end{bmatrix}=\begin{bmatrix} 6 & -3 \\ 4 & 7 \end{bmatrix}^{-1}\begin{bmatrix} 39 \\ 539 \end{bmatrix}=\begin{bmatrix} 35 \\ 57 \end{bmatrix}$$

3. (−18, −29, 75)

$$\begin{bmatrix} 8 & 5 & 6 \\ 3 & -4 & -5 \\ -4 & -13 & 7 \end{bmatrix}\begin{bmatrix} x \\ y \\ z \end{bmatrix}=\begin{bmatrix} 161 \\ -313 \\ 974 \end{bmatrix}$$

$$\begin{bmatrix} x \\ y \\ z \end{bmatrix}=\begin{bmatrix} 8 & 5 & 6 \\ 3 & -4 & -5 \\ -4 & -13 & 7 \end{bmatrix}^{-1}\begin{bmatrix} 161 \\ -313 \\ 974 \end{bmatrix}=\begin{bmatrix} -18 \\ -29 \\ 75 \end{bmatrix}$$

4. (5, −7, 9, −11)

$$\begin{bmatrix} 5 & -3 & 6 & -7 \\ 8 & 12 & 6 & -5 \\ 6 & 3 & -5 & 3 \\ 1 & -1 & 1 & -1 \end{bmatrix}\begin{bmatrix} w \\ x \\ y \\ z \end{bmatrix}=\begin{bmatrix} 177 \\ 65 \\ -69 \\ 32 \end{bmatrix}$$

$$\begin{bmatrix} w \\ x \\ y \\ z \end{bmatrix}=\begin{bmatrix} 5 & -3 & 6 & -7 \\ 8 & 12 & 6 & -5 \\ 6 & 3 & -5 & 3 \\ 1 & -1 & 1 & -1 \end{bmatrix}^{-1}\begin{bmatrix} 177 \\ 65 \\ -69 \\ 32 \end{bmatrix}=\begin{bmatrix} 5 \\ -7 \\ 9 \\ -11 \end{bmatrix}$$

5.

Tickets Sold	Student	Adult
In Advance	450	750
At the Door	350	400

t: number of student tickets sold in advance

d: number of student tickets sold at the door

a: number of adult tickets sold in advance

b: number of adult tickets sold at the door

$t+d+a+b=1{,}950$

$8t+10d+12a+15b=22{,}100$

$a+b=t+d+350$

$12a+15b=8t+10d+7{,}900$

Rewrite the equations with all variable terms on the left:

$t+d+a+b=1{,}950$

$8t+10d+12a+15b=22{,}100$

$-t-d+a+b=350$

$-8t-10d+12a+15b=7{,}900$

Matrix equation and solution

$$\begin{bmatrix} 1 & 1 & 1 & 1 \\ 8 & 10 & 12 & 15 \\ -1 & -1 & 1 & 1 \\ -8 & -10 & 12 & 15 \end{bmatrix}\begin{bmatrix} t \\ d \\ a \\ b \end{bmatrix}=\begin{bmatrix} 1{,}950 \\ 22{,}100 \\ 350 \\ 7{,}900 \end{bmatrix}$$

$$\begin{bmatrix} t \\ d \\ a \\ b \end{bmatrix}=\begin{bmatrix} 1 & 1 & 1 & 1 \\ 8 & 10 & 12 & 15 \\ -1 & -1 & 1 & 1 \\ -8 & -10 & 12 & 15 \end{bmatrix}^{-1}\begin{bmatrix} 1{,}950 \\ 22{,}100 \\ 350 \\ 7{,}900 \end{bmatrix}=\begin{bmatrix} 450 \\ 350 \\ 750 \\ 400 \end{bmatrix}$$

EXERCISE 1.7

1.
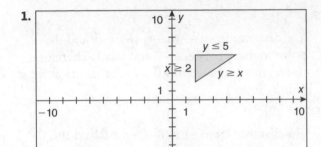

2.
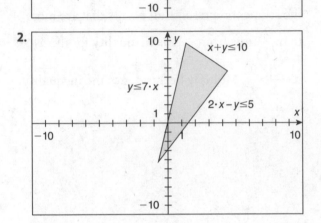

3.
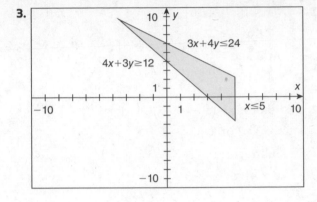

EXERCISE 1.8

1. –7, –3

$x+5=\pm2$

$x+5=2$ or $x+5=-2$

$x=-3,-7$

2. 8, 22

$x-15=\pm7$

$x-15=7$ or $x-15=-7$

$x=22,8$

3. –2, 5

$2x-3=\pm7$

$2x-3=7$ or $2x-3=-7$

$2x=10$ or $2x=-4$

$x=5,-2$

4. –0.8, 2

$5x-3=\pm7$

$5x-3=7$ or $5x-3=-7$

$5x=10$ or $5x=-4$

$x=2,\dfrac{-4}{5}$

5. –2.5, 8

$11-4x=\pm21$

$11-4x=21$ or $11-4x=-21$

$-4x=10$ or $-4x=-32$

$x=-2.5,8$

6. 11, 20

$|31-2x|=|2x-31|$

$2x-31=\pm9$

$2x-31=9$ or $2x-31=-9$

$2x=40$ or $2x=22$

$x=20,11$

EXERCISE 1.9

1. $-4 \le x \le 10; \left[-4, 10\right]$

$-7 \le x - 3 \le 7$

Add 3:

$-4 \le 10$

2. $x < -5$ or $x > -1; \left(-\infty, -5\right) \cup \left(-1, \infty\right)$

$x + 3 > 2$ or $x + 3 < -2$

Subtract 3:

$x > -1$ or $x < -5$

3. $x \le -1$ or $x \ge 2.5; \left(-\infty, -1\right] \cup \left[2.5, \infty\right)$

$4x - 3 \ge 7$ or $4x - 3 \le -7$

Add 3:

$4x \ge 10$ or $4x \le -4$

$x \ge 2.5$ or $x \le -1$

4. $1 < x < \dfrac{11}{3}; \left(1, \dfrac{11}{3}\right)$

$\left|7 - 3x\right| = \left|3x - 7\right|$

$-4 < 3x - 7 < 4$

Add 7:

$3 < 3x < 11$

Divide by 3:

$1 < x < \dfrac{11}{3}$

5. $\left|x - 1\right| \le 8$

The distance from –7 to 9 is 16 units and the point midway between –7 and 9 is 1. Therefore, the segment represents all points that are at most 8 units from 1.

6. $\left|2x - 7\right| > 9$

The distance from –1 to 8 is 9 units and the point midway between –1 and 8 is $\dfrac{7}{2}$. Therefore, the segment represents all points that are more than $\dfrac{9}{2}$ units from $\dfrac{7}{2}$. The inequality for this is $\left|x - \dfrac{7}{2}\right| > \dfrac{9}{2}$. Multiply both sides of the inequality by 2.

CHAPTER 2

Functions

EXERCISE 2.1

1. {−1, 3, 4, 6}

All the first elements in relation A. The element 4 does not have to be written twice.

2. {0, 3, 7, 10, 12}

All the second elements in relation A.

3. {0, 3, 5, 7, 10}

4. {−10, −2, −1, 1, 5}

5. {Kristen, Stacey, Kate, Colin, Carson, Brendon, Russ, Andrew}

6. {5, 8, 9, 12, 15, 17, 21}

7. {(7, −1), (3, 6), (0, 4), (10, 4), (12, 3)}

Interchange all the first and second elements in relation A.

8. {(−1, 10), (−2, 0), (−10, 7), (5, 3), (−2, 7), (1, 5)}

9. {(5, Kristen), (21, Stacey), (9, Kate), (8, Colin), (12, Carson), (15, Brendon), (12, Russ), (17, Andrew)}

EXERCISE 2.2

1. C

Each element in the domain of C is matched with only one element in the range of C.

2. B^{-1}, C^{-1}

A^{-1} contains the elements (2, −5) and (2, 13) so A^{-1} is not a function.

3. No, as rare as it might be, students may share the same e-mail.

4. No, a student may have multiple e-mails.

5. Yes, a student may only have one Social Security number (at least legally).

6. Yes, Social Security numbers are not (legally) shared.

7. Yes, each person has only one biological mother.

8. No, siblings in the same homeroom could have the same biological mother.

EXERCISE 2.3

1. −16

$$f(4) = -7(4) + 12 = -28 + 12 = -16$$

2. $-7n -16$

$$f(n + 4) = -7(n + 4) + 12 = -7n - 28 + 12 = -7n - 16$$

3. $\dfrac{31}{5}$

$$g(7) = \frac{4(7)+3}{7-2} = \frac{31}{5}$$

4. $\dfrac{4t+11}{t}$

$$g(t+2) = \frac{4(t+2)+3}{t+2-2} = \frac{4t+8+3}{t} = \frac{4t+11}{t}$$

5. 7

$$p(5) = \sqrt{8(5)+9} = \sqrt{49} = 7$$

6. $\sqrt{8r-15}$

$$p(r-3) = \sqrt{8(r-3)+9} = \sqrt{8r-15}$$

EXERCISE 2.4

1. 24

$$f(-2)+g(3)=2(-2)^2-3(-2)+5+\sqrt{5(3)+10}=8+6+5+\sqrt{25}=24$$

2. $3\sqrt{5}$

$$f(2)=2(2)^2-3(2)+5=8-6+5=7$$

$$g(f(2))=g(7)=\sqrt{5(7)+10}=\sqrt{45}=\sqrt{9*5}=3\sqrt{5}$$

3. $14\sqrt{10}$

$$f(3)*g(0)=\left(2(3)^2-3(3)+5\right)\left(\sqrt{5(0)+10}\right)=(14)\left(\sqrt{10}\right)$$

4. $\dfrac{35}{3}$

$$\frac{f(10)}{g(43)}=\frac{2(10)^2-3(10)+5}{\sqrt{5(43)+10}}=\frac{200-30+5}{\sqrt{215+10}}=\frac{175}{\sqrt{225}}=\frac{175}{15}=\frac{35}{3}$$

5. 1,180

$$g(123)=\sqrt{5(123)+10}=\sqrt{615+10}=\sqrt{625}=25$$

$$f(g(123))=f(25)=2(25)^2-3(25)+5=1,250-75+5=1,180$$

6. $5\sqrt{7}-19$

$$g(33)-f(-2)=\sqrt{5(33)+10}-(19)=\sqrt{175}-19=\sqrt{25*7}-19=5\sqrt{7}-19$$

7. $x \neq \dfrac{4}{3}$

Set the denominator equal to 0.

8. $x \neq \pm 2$

Set the denominator equal to 0.

9. $x \geq -9$

Set the radicand ≥ 0.

10. $x \leq 9$

Set the radicand ≥ 0.

11. 100

Breakeven occurs when cost = revenue. Revenue is the product of the number of units sold and the price per unit, so $R(n) = 400n$. Solve $400n = 150n + 25,000$.

EXERCISE 2.5

1. Left 1, reflect over the x-axis, dilate from the x axis by a factor of $\dfrac{1}{2}$, down 3.

Work from the inside out: $x + 1$: moves the graph left 1; $(x + 1)^2$: base function; $-(x + 1)^2$: reflect the graph over the x-axis; $\dfrac{-1}{2}(x+1)^2$: compress the graph to the x-axis by a factor of $\dfrac{1}{2}$; $\dfrac{-1}{2}(x+1)^2 - 3$: move the graph down 3.

2. Right 4, down 1

3. Left 9, up 1

4. Reflect over the x axis, dilate from the x axis by a factor of 4, up 9

5. Left 2, reflect over the x axis, dilate from the x axis by a factor of 2 , up 5

6. Left 2, dilate from the x-axis by a factor of 2, down 3

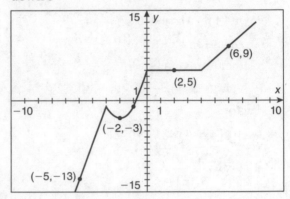

EXERCISE 2.6

1. $f^{-1}(x) = \dfrac{5-x}{2} = \dfrac{-x+5}{2}$

$\quad f: y = -2x + 5$

$\quad f^{-1}: x = -2y + 5$

$\quad\quad x - 5 = -2y$

$\quad f^{-1}(x) = \dfrac{x-5}{-2} = \dfrac{5-x}{2}$

2. $g^{-1}(x) = \dfrac{x-4}{-5}$

$\quad g: y = 4 - 5x$

$\quad g^{-1}: x = 4 - 5y$

$\quad\quad x - 4 = -5y$

$\quad g^{-1}(x) = \dfrac{x-4}{-5} = \dfrac{4-x}{5}$

3. $k^{-1}(x) = \dfrac{-7x-2}{x-5}$

$\quad k: \ y = \dfrac{5x-2}{x+7}$

$\quad k^{-1}: x = \dfrac{5y-2}{y+7}$

$\quad\quad x(y+7) = 5y - 2$

$\quad\quad xy + 7x = 5y - 2$

Gather terms in y:

$\quad\quad xy - 5y = -7x - 2$

Factor:

$\quad\quad y(x-5) = -7x - 2$

$\quad k^{-1}(x) = \dfrac{-7x-2}{x-5}$

4. $v^{-1}(x) = \dfrac{3x-10}{4x-1}$

$\quad v:\ y = \dfrac{x-10}{4x-3}$

$\quad v^{-1}:\ x = \dfrac{y-10}{4y-3}$

$\qquad x(4y-3) = y-10$

$\qquad 4xy - 3x = y - 10$

Gather terms in y:

$\qquad 4xy - y = 3x - 10$

Factor:

$\qquad y(4x-1) = 3x - 10$

$\quad v^{-1}(x) = \dfrac{3x-10}{4x-1}$

EXERCISE 2.7

1. (a), (d)

Passes the vertical line test.

2. (b), (c), (d)

Passes the horizontal line test.

3. (b)

(c) clearly fails the vertical line test; (a) is a parabola opening sideways; (b) is a function because each element in the domain has a unique element in the range.

4. (a)

(a) is a left opening parabola, so its inverse will be a downward opening parabola; the inverse of (b) contains the ordered pairs (3, 2) and (3, 7); (c) fails the horizontal line test.

CHAPTER 3

Quadratic Relationships

EXERCISE 3.1

1. $(2x - 7)^2 = (2x)^2 - 2(2x)(7) + (7)^2$
2. $(3x + 5y)^2 = (3x)^2 + 2(3x)(5y) + (5y)^2$
3. $(10x - 7y)(10x + 7y)$

4. $9(x + 5y)^2 = 9\left((x)^2 + 2(x)(5y) + (5y)^2\right)$
5. $(5x - 8z^2)(25x^2 + 40xz^2 + 64z^4) = (5x)^3 - (8z)^3$
6. $5(3c + 2)(9c^2 - 6c + 4) = 5\left((3c)^2 + (2)^3\right)$

EXERCISE 3.2

1. $(2x - 3)(3x - 2)$
2. $(4x - 3)(5x + 4)$
3. $(2x + 9)(6x - 5)$

4. $10(3x - 4)(2x - 5) = 10(6x^2 - 23x + 20)$
5. $(13x - 4)(2x + 5)$

EXERCISE 3.3

1. $2(x + 3)^2 - 13$

   $2\left(x^2 + 6x + \quad\right) + 5$

   $2\left(x^2 + 6x + 3^2\right) + 5 - 2(3)^2$

2. $2\left(x - \dfrac{7}{4}\right)^2 - \dfrac{17}{8}$

   $2\left(x^2 - \dfrac{7}{2}x + \quad\right) + 4$

   $2\left(x^2 - \dfrac{7}{2}x + \left(\dfrac{7}{4}\right)^2\right) + 5 - 2\left(\dfrac{7}{4}\right)^2$

3. $-5(x - 2)^2 + 21$

   $-5\left(x^2 - 4x + \quad\right) + 1$

   $-5\left(x^2 - 4x + (2)^2\right) + 1 + 5(2)^2$

4. $\dfrac{1}{2}(x - 3)^2 - \dfrac{27}{2}$

   $\dfrac{-1}{2}\left(x^2 - 6x + \quad\right) - 9$

   $\dfrac{-1}{2}\left(x^2 - 6x + (3)^2\right) - 9 + \dfrac{1}{2}(3)^2$

5. $\dfrac{2}{3}(x - 9)^2 - 65$

   $\dfrac{2}{3}\left(x^2 - 18x + \quad\right) - 11$

   $\dfrac{2}{3}\left(x^2 - 18x + (9)^2\right) - 11 - \dfrac{2}{3}(9)^2$

EXERCISE 3.4

1. $\dfrac{11}{18} \pm \dfrac{\sqrt{193}}{18}$

$b^2 - 4ac = (-11)^2 - 4(9)(-2) = 121 + 72 = 193$

2. $\dfrac{27}{28} \pm \dfrac{\sqrt{1,345}}{28}$

$b^2 - 4ac = (27)^2 - 4(-14)(11) = 7,219 + 616 = 1,345$

3. $\dfrac{-1}{4}, 6$

$b^2 - 4ac = (25)^2 - 4(4)(6) = 625 - 96 = 529$

4. $\dfrac{-25}{8} \pm \dfrac{\sqrt{721}}{8}$

$b^2 - 4ac = (25)^2 - 4(4)(-6) = 625 + 96 = 721$

5. $\dfrac{-13}{24} \pm \dfrac{\sqrt{313}}{24}$

$b^2 - 4ac = (13)^2 - 4(12)(-3) = 169 + 144 = 313$

EXERCISE 3.5

1. $x = -2$, $(-2, -16)$

$\dfrac{-b}{2a} = \dfrac{-12}{2(3)} = -2$

2. $x = -5$, $(-5, 2)$

$h = -5$

3. $x = \dfrac{3}{4}$, $\left(\dfrac{3}{4}, \dfrac{41}{8}\right)$

$\dfrac{-b}{2a} = \dfrac{-3}{2(-2)} = \dfrac{3}{4}$

4. $x = 3$, $(3, 6)$

$h = 3$

5. $x = 6$, $(6, 10)$

$\dfrac{-b}{2a} = \dfrac{\frac{-8}{3}}{2\left(\frac{-2}{9}\right)} = 6$

EXERCISE 3.6

1. 17.86 ft by 31.25 ft.

   L: length of fence along the shore line

   w: length of fence perpendicular to the shore line

   $28L + 2(8w) = 1,000$ so $L = \dfrac{1,000 - 16w}{28}$

   $A = Lw$ or $A = \left(\dfrac{1,000 - 16w}{28}\right) w = \dfrac{250}{7} w - \dfrac{4}{7} w^2$

2. $1,550

   $R(n) = (150 - 0.5n)n$

   $P(n) = R(n) - C(n) = -0.5n^2 + 70n - 900$

3. 276 ft

   Ball reaches maximum height when

   $t = \dfrac{-112}{2(-16)} = \dfrac{7}{2}$ seconds

   Maximum height is $-16(1.75)^2 + 112(1.75) + 80$

4. 48 ft/sec

   Average speed $= \dfrac{h(3) - h(1)}{3 - 1}$

5. 7.65 seconds

   Solve the equation $h = 0$.

6. 121.8 ft

Ball reaches maximum height when

$t = \dfrac{-88.3}{2(-16)} = 2.759$ seconds

Maximum height is $-16(2.759)^2 + 88.3(2.759)$ feet

7. 5.52 seconds

Ball strikes the ground when $v = 0$. Solve $-16t^2 + 88.3t = 0$

8. 235.4 yds

Distance in feet is $117.1(5.52)$. Divide by 3 and add 20 for the roll.

EXERCISE 3.7

1. $(6x - 5)(12x + 17)$

x-intercepts: $-1.42, 0.8333$ (both results rounded)

2. $(8x - 7)(9x - 14)$

x-intercepts: $0.875, 1.56$ (result rounded)

3. $(15x + 8)(12x - 7)$

x-intercepts: $-0.533, 0.583$ (both results rounded)

4. $(25x - 19)(14x + 9)$

x-intercepts: $0.76, -0.65$ (result rounded)

5. $(3x - 19)(4x + 17)$

x-intercepts: $-4.25, 6.33$ (result rounded)

EXERCISE 3.8

1. $y = \dfrac{-33}{32}, \left(-2, \dfrac{-31}{32}\right)$

$\dfrac{1}{4p} = 8 \Rightarrow p = \dfrac{1}{32}$

2. $y = \dfrac{-1}{16}(x-2)^2 + 9$

$p = 4$ (distance from focus to vertex)

parabola is concave down

3. $y = \dfrac{-1}{12}(x-2)^2 + 8$

$2p = 6$ (distance from focus to directrix)

parabola is concave down

4. $x = \dfrac{1}{24}(y-3)^2 + 2$

$2p = 12$ (distance from focus to directrix)

parabola opens to the right

5. $x = 7$

$\dfrac{1}{4p} = \dfrac{-1}{32} \Rightarrow p = -8$

Vertex at $(-1, 8)$; parabola opens to the left

Directrix 8 units to the right of the vertex

EXERCISE 3.9

1. $(x+4)(x+5)(x-5)$

$x^2(x+4) - 25(x+4)$

2. $(x+4)(x-5)(x^2+5x+25)$

$x^3(x+4) - 125(x+4)$

3. $(x-9)(x+9)(x-7)$

$x^2(x-7) - 81(x-7)$

4. $(x+3)^2(x^2-3x+9)$

$x^3(x+3) + 27(x+3)$

5. $(x-5)(x^2+27)$

$x^2(x-5) + 27(x-5)$

EXERCISE 3.10

1. $5x(x-9)(x+9)$

$5x$ is a common factor

2. $(2x-9y)^2$

$(2x)^2 - 2(2x)(9y) + (9y)^2$

3. $(5x+3)(4x-9)$

$(9)(5) - (3)(4) = 33$

4. $x(14x-9)(11x+12)$

x is a common factor

x intercepts: 0, –1.09, 0.643 (both results rounded)

5. $(x-8)(x-4)(x+4)$

$x^2(x-8) - 16(x-8)$

6. $(x^2-5)^2$

$(x^2)^2 - 2(x^2)(5) + (5)^2$

7. $(x-5)^2(x+5)^2$

$(x^2)^2 - 2(x^2)(25) + (25)^2$

8. $(x+4)(x-10)(x^2+10x+100)$

$x^3(x+4) - 1{,}000(x+4)$

9. $(3x-5)(5x-3)$

$5^2 + 3^2 = 34$

10. $x(3x+5)(16x-9)$

x intercepts: 0, –1.67, 0.563 (both results rounded)

EXERCISE 3.11

1. $C\ (6, -4)$; $r = 14$

Complete the square to get $(x-6)^2 + (y+4)^2 = 196$

2. $(x+4)^2 + (y-5)^2 = 25$

Distance from (–4, 5) to x-axis is 5

3. $(x-2)^2 + (y-16)^2 = 61$

Center is at the midpoint: (2, 16)

Distance from center to either endpoint is $\sqrt{61}$

4. $\left(\pm\sqrt{41}, 2\right)$

Substitute $x^2 = 45 - y^2$ into second equation

$45 - y^2 + y^2 + 6y + 9 = 66$

$6y = 12$

$y = 2$

5. (–5, 0), (0, 5), (–4, 3), (3, –4)

Substitute for y in the circle equation:

$x^2 + \left(\dfrac{-1}{2}x^2 - \dfrac{3}{2}x + 5\right)^2 = 25$

$x^2 + \dfrac{1}{4}x^4 + \dfrac{3}{2}x^3 - \dfrac{11}{4}x^2 - 15x + 25 = 25$

$\dfrac{1}{4}x^4 + \dfrac{3}{2}x^3 - \dfrac{7}{4}x^2 - 15x = 0$

Multiply by 4:

$x^4 + 6x^3 - 7x^2 - 60x = 0$

Graph:

$x\text{-intercepts:} 0, -5, -4, 3$

CHAPTER 4

Complex Numbers

EXERCISE 4.1

1. -1

$$i^{38} = i^2$$

2. -1

$$i^{138} = i^2$$

3. $-i$

$$i^{57} = i^3$$

EXERCISE 4.2

1. $-4i\sqrt{2}$

$$4\sqrt{-1}\sqrt{25}\sqrt{2} - 6\sqrt{-1}\sqrt{16}\sqrt{2}$$

2. $49i\sqrt{6}$

$$17\sqrt{-1}\sqrt{4}\sqrt{6} + 5\sqrt{-1}\sqrt{9}\sqrt{6}$$

3. -72

$$\left(\sqrt{-1}\right)^2\left(\sqrt{72}\right)^2$$

4. $-960\sqrt{15}$

$$\left(12i\sqrt{4}\sqrt{5}\right)\left(20i\sqrt{4}\sqrt{3}\right)$$

5. $2\sqrt{6}$

$$\frac{52i\sqrt{9}\sqrt{2}}{13i\sqrt{4}\sqrt{3}}$$

6. $-18{,}000a$

$$\left(40i\sqrt{9}\sqrt{5a}\right)\left(15i\sqrt{4}\sqrt{5a}\right)$$

7. $4a$

$$\frac{40ai\sqrt{9}\sqrt{5a}}{15i\sqrt{4}\sqrt{5a}}$$

EXERCISE 4.3

1. $8a + 2i\sqrt{2b}$

$$(a+7a) + \left(-2\sqrt{2b} + 4\sqrt{2b}\right)i$$

2. $7a^2 + 16b - 10ai\sqrt{2b}$

$$7a^2 + a\left(\sqrt{-1}\sqrt{16}\sqrt{2b}\right) - 7a\left(\sqrt{-1}\sqrt{4}\sqrt{2b}\right) - \left(\sqrt{-1}\sqrt{16}\sqrt{2b}\right)\left(\sqrt{-1}\sqrt{4}\sqrt{2b}\right)$$

3. $\dfrac{7a^2 - 16b}{49a^2 + 32b} - \dfrac{18a\sqrt{2b}}{49a^2 + 32b}i$

$$\left(\frac{a - \sqrt{-8b}}{7a + \sqrt{-32b}}\right)\left(\frac{7a - \sqrt{-32b}}{7a - \sqrt{-32b}}\right)$$

$$\frac{7a^2 - 4ai\sqrt{2b} - 14ai\sqrt{2b} - 16b}{49a^2 + 32b}$$

4. $\dfrac{-11}{13} - \dfrac{4\sqrt{3}}{13}i$

$$\left(\dfrac{4-8i\sqrt{3}}{4+8i\sqrt{3}}\right)\left(\dfrac{4-8i\sqrt{3}}{4-8i\sqrt{3}}\right)$$

$$\dfrac{16-64i\sqrt{3}-192}{16+192}$$

$$\dfrac{-176}{208} - \dfrac{64\sqrt{3}}{208}i$$

EXERCISE 4.4

1. Real, irrational, unequal

   Discriminant = $49 - 4(9)(-3) = 157$

2. Real, rational, equal (double root)

   Discriminant = $576 - 4(9)(16) = 0$

3. Complex conjugates

   Discriminant = $144 - 4(9)(16) = -432$

4. $a < \dfrac{16}{5}$

   Solve: $64 - 4a(5) > 0$

5. 0, 20

   Solve: $b^2 - 20b = 0$

EXERCISE 4.5

1. $\dfrac{-5}{2}$

   $$S = \dfrac{-b}{a} = \dfrac{-25}{10}$$

2. $\dfrac{-17}{10}$

   $$P = \dfrac{c}{a} = \dfrac{-17}{10}$$

3. $50x^2 + 5x - 21 = 0$

   $$P = \dfrac{-21}{50}; S = \dfrac{3}{5} + \dfrac{-7}{10} = \dfrac{-1}{10} = \dfrac{-5}{50}$$

   $a = 50; b = 5; c = -21$

4. $x^2 - 10x + 13 = 0$

   $P = 13; S = 10$

   $a = 1; b = -10; c = 13$

5. $192x^2 - 240x + 271 = 0$

   $$P = \dfrac{25}{64} + \dfrac{147}{144} = \dfrac{271}{192}; S = \dfrac{10}{4} = \dfrac{480}{192}$$

   $a = 192; b = -240; c = 271$

CHAPTER 5

Polynomial Functions

EXERCISE 5.1

1. Even

Symmetric to the y axis

2. Odd

Symmetric to the origin

3. Neither

4. Even

Symmetric to the y axis

EXERCISE 5.2

1. $x \to \pm\infty \Rightarrow f(x) \to -\infty$

Largest exponent is even and leading coefficient is negative

2. $x \to -\infty \Rightarrow f(x) \to -\infty$ and $x \to \infty \Rightarrow f(x) \to \infty$

Largest exponent is odd and leading coefficient is positive

3. $x \to -\infty \Rightarrow f(x) \to -\infty$ and $x \to \infty \Rightarrow f(x) \to \infty$

Don't get fooled, the largest exponent is a 9.

4. $x \to \pm\infty \Rightarrow f(x) \to \infty$

Largest exponent is even and leading coefficient is positive

EXERCISE 5.3

1. Yes

$P(-3) = 0$

2. Yes

$P(5) = 0$

3. Yes

$P\left(\dfrac{3}{2}\right) = 0$

4. Yes

$P(-6) = 0$

5. No

$P(6) = 1,620$

6. No

$P(4) = 1,904$

7. Yes

$P\left(\dfrac{5}{3}\right) = 0$

8. Yes

$P\left(\dfrac{5}{2}\right) = 0$

EXERCISE 5.4

1. $16x^2 - 1$

$-3|$	16	48	-1	-3
		-48	0	3
	16	0	-1	0

2. $6x^2 - 7x - 3$

$5|$	6	-37	32	15
		30	-35	-15
	6	-7	-3	0

3. $3x^2 - 14x - 5$

$\dfrac{3}{2}|$	6	-37	32	15
		9	-42	-15
	6	-28	-10	0

$$\left(6x^3 - 37x^2 + 32x + 15\right) \div \left(2x - 3\right) \text{ becomes}$$

$$\left(6x^3 - 37x^2 + 32x + 15\right) \div 2\left(x - \frac{3}{2}\right)$$

$$\left(6x^2 - 28x - 10\right) \div 2 = 3x^2 - 14x - 5$$

4. $4x^3 - 20x^2 - 9x + 45$

$-6|$	4	4	-129	-9	270
		-24	120	54	-270
	4	-20	-9	45	0

5. $4x^3 + 28x^2 + 39x + 225 + \dfrac{1,620}{x-6}$

$6|$	4	4	-129	-9	270
		24	168	234	1,350
	4	28	39	225	1,620

6. $9x^3 + 54x^2 + 119x + 426 + \dfrac{1,904}{x-4}$

$4|$	9	18	-97	-50	200
		36	216	476	1,704
	9	54	119	426	1,904

7. $3x^3 + 11x^2 - 14x - 40$

$\dfrac{5}{3}|$	9	18	-97	-50	200
		15	55	-70	-200
	9	33	-42	-120	0

$$\left(9x^3 + 33x^2 - 42x - 120\right) \div 3 = 3x^3 + 11x^2 - 14x - 40$$

8. $x^3 + 8x^2 + 22x + 24$

$\dfrac{5}{2}|$	2	11	4	-62	120
		5	40	110	120
	2	16	44	48	0

$$\left(2x^3 + 16x^2 + 44x + 48\right) \div 2 = x^3 + 8x^2 + 22x + 24$$

9. $x^4 + 2x^3 - 4x^2 + 5x - 5$

$-2|$	1	4	0	-3	5	-10
		-2	-4	8	-10	10
	1	2	-4	5	-5	0

10. $\left(x+3\right)\left(4x-1\right)\left(4x+1\right)$

x intercepts: -3, $-.25$, $.25$

11. $\left(x-5\right)\left(2x-3\right)\left(3x+1\right)$

x intercepts: 5, $-.333$, 1.5 (answer rounded)

12. $\left(x+4\right)\left(3x-5\right)\left(3x+5\right)\left(x-2\right)$

x intercepts: 2, -4, 1.667, 1.667 (both answers rounded)

13. $\left(x-5\right)\left(2x-3\right)\left(2x+3\right)\left(x+6\right)$

x intercepts: 5, 1.5, -1.5, -6

14. $\left(x-3\right)\left(x^2+3x+4\right)$

x intercept: $x = 3$

EXERCISE 5.5

1. $\pm 1.5, 2, \dfrac{-3}{2} \pm \dfrac{\sqrt{7}}{2}i$

x intercepts: 1.5, –1.5, 2

Reduce polynomial:

$2|$	4	4	–17	–41	18	72
		8	24	14	–54	–72
	4	12	7	–27	–36	0

$\dfrac{3}{2}|$	4	12	7	–27	–36
		6	27	51	36
	4	18	7	24	0

$\dfrac{-3}{2}|$	4	18	34	24
		–6	–18	–24
	4	12	7	0

$4x^5 + 4x^4 - 17x^3 - 41x^2 + 18x + 72 =$

$4(x-2)\left(x-\dfrac{3}{2}\right)\left(x+\dfrac{3}{2}\right)(x^2+3x+4)$

Solve $x^2 + 3x + 4 = 0$.

2. $\dfrac{2}{3}, \dfrac{3}{2}, \pm 2i$

x intercepts: 1.5, 0.667 (answer reduced)

Reduce polynomial:

$\dfrac{3}{2}|$	6	–13	30	–52	24
		9	–6	36	–24
	6	–4	24	–16	0

$\dfrac{2}{3}|$	6	–4	24	–16
		4	0	16
	6	0	24	0

3. $\dfrac{-7}{4}, \dfrac{5}{6}, \dfrac{-5}{4} \pm \dfrac{\sqrt{31}}{4}i$

x-intercepts: –1.75, 0.833 (answer rounded)

$\dfrac{-7}{4}|$	48	164	208	–21	–245
		–84	–140	–119	245
	48	80	68	–140	0

$\dfrac{5}{6}|$	48	80	68	–140
		40	100	140
	48	120	168	0

$48x^4 + 164x^3 + 208x^2 - 21x - 245 =$

$(4x+7)(6x-5)(2x^2+5x+7)$

$2x^2 + 5x + 7 = 0 \Rightarrow x = \dfrac{-5}{4} \pm \dfrac{\sqrt{25-4(2)(7)}}{4} =$

$\dfrac{-5}{4} \pm \dfrac{\sqrt{-31}}{4} = \dfrac{-5}{4} \pm \dfrac{\sqrt{31}}{4}i$

4. $\dfrac{-3}{4}, \dfrac{7}{3}, 1 \pm i\sqrt{2}$

x-intercepts: –0.75, 2.333 (answer reduced)

$\dfrac{-3}{4}|$	48	–172	212	–60	–252
		–36	156	–276	252
	48	–208	368	–336	0

$\dfrac{7}{3}|$	48	–208	368	–336
		112	–224	336
	48	–96	144	0

$48x^4 - 172x^3 + 212x^2 - 60x - 252 =$

$4(4x+3)(3x-7)(x^2-2x+3)$

$x^2 - 2x + 3 = 0 \Rightarrow x = \dfrac{2}{2} \pm \dfrac{\sqrt{4-12}}{2} = 1 \pm 2i\sqrt{2}$

5. $\pm 2, \pm 5, \pm 2i$

$x^4(x^2-25) - 16(x^2-25) = 0$

$(x^2-25)(x^4-16) = 0$

$(x-5)(x+5)(x^2-4)(x^2+4) = 0$

EXERCISE 5.6

1. $0 < x < 1 \cup x > 8$

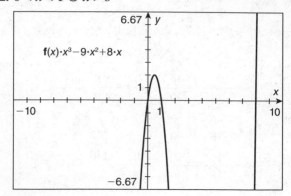

2. $\dfrac{-12}{5} \le x \le \dfrac{3}{4} \cup x \ge 6$

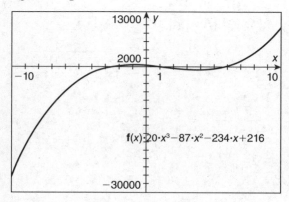

3. $x \ne 4$

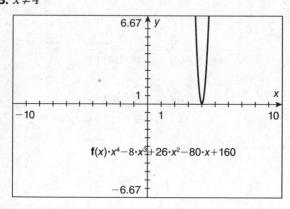

4. $x \le -4 \cup 2 \le x \le 4$

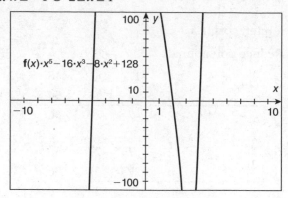

5. $-2 < x < \dfrac{-5}{3} \cup -1 < x < 1 \cup \dfrac{5}{3} < x < 2$

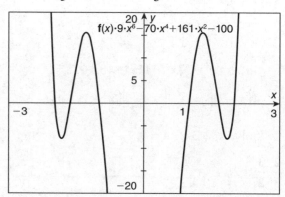

CHAPTER 6

Rational and Irrational Functions

EXERCISE 6.1

1. $x \neq -2$

2. $x \neq -1, \dfrac{1}{2}$

3. $x \neq \dfrac{2}{3}, 1$

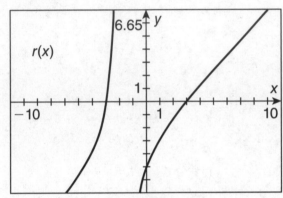

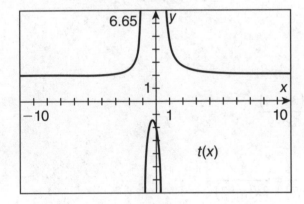

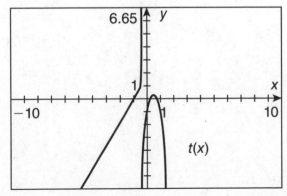

4. $x \to -\infty, y \to -\infty;\ x \to \infty, y \to \infty$

5. $x \to \pm\infty, y \to 2$

6. $x \to -\infty, y \to -\infty;\ x \to \infty, y \to \infty$

7. All real numbers

8. $y \leq -1.3 \cup y > 2$

9. $y \leq 0.343 \cup y \geq 11.9 \cup y \neq \dfrac{-20}{3}$

EXERCISE 6.2

1. $\dfrac{(2x-3)\cancel{(3x+5)}}{(4x-5)\cancel{(5x+4)}} \cdot \dfrac{\cancel{(5x+4)}\cancel{(3x-2)}}{\cancel{(3x+5)}\cancel{(3x-2)}} = \dfrac{2x-3}{4x-5}$

2. $\dfrac{\cancel{(6x+5)}\cancel{(6x-5)}}{3\cancel{(3x+4)}\cancel{(6x+5)}} \cdot \dfrac{2\cancel{(2x-7)}\cancel{(3x+4)}}{\cancel{(2x-7)}\cancel{(6x-5)}} = \dfrac{2}{3}$

3. $\dfrac{\cancel{(4x-7)}(3x-8)}{\cancel{(4x+5)}(2x-3)} \cdot \dfrac{(x+5)\cancel{(4x+5)}}{-(2x+9)\cancel{(4x-7)}} = \dfrac{-(3x-8)(x+5)}{(2x-3)(2x+9)}$

4. $\dfrac{\cancel{(x-6)}\cancel{(x^2+6x+36)}}{5\cancel{(x-6)}(x+6)} \cdot \dfrac{7(x-3)\cancel{(x+6)}}{4\cancel{(x^2+6x+36)}} = \dfrac{7(x-3)}{20}$

5. $\dfrac{\cancel{(x-3)}\cancel{(x+3)}\cancel{(x^2+9)}}{\cancel{(x^2+9)}\cancel{(x-3)}} \cdot \dfrac{8\cancel{(x^2-3x+9)}}{\cancel{(x+3)}\cancel{(x^2-3x+9)}} = 8$

6. $\dfrac{(2x-3)(2x+3)(x-5)}{(2x-7)\cancel{(3x+2)}} \cdot \dfrac{\cancel{(3x+2)}\cancel{(4x^2+9)}}{(2x-3)(2x+3)\cancel{(4x^2+9)}} = \dfrac{x-5}{2x-7}$

7. $\dfrac{(3x+4)(4x+3)}{(5x+2)(2x+5)} \times \dfrac{\cancel{(7x+3)}(3x+4)}{\cancel{(7x+3)}(5x+2)} = \dfrac{(4x+3)(3x+4)^2}{(2x+5)(5x+2)^2}$

8. $\dfrac{x\cancel{(4x-1)}\cancel{(2x+3)}}{6\cancel{(2x+3)}(3x+2)} \times \dfrac{\cancel{2}x(x+5)\cancel{(3x+2)}}{\cancel{2}\cancel{(4x-1)}(x+3)} = \dfrac{x^2(x+5)}{6(x+3)}$

9. $\dfrac{(x-2)(2x+3)\cancel{(4x-9)}}{\cancel{(4x-9)}\cancel{(4x+9)}} \times \dfrac{(3x-4)\cancel{(4x+9)}}{(2x+3)(4x^2-6x+9)} = \dfrac{(x-2)(3x-4)}{4x^2-6x+9}$

10. $\dfrac{3x\cancel{(2x+3)}\cancel{(9x-8)}}{\cancel{(2x+3)}\cancel{(9x-8)}(9x+8)} \times \dfrac{\cancel{(8x+9)}(9x+8)}{6x^2(x+7)\cancel{(8x+9)}} = \dfrac{3x}{6x^2(x+7)} = \dfrac{1}{2x(x+7)}$

EXERCISE 6.3

1. $\dfrac{(2)(2)(x-1)}{2(3)(x+1)} - \dfrac{3(1)(x+1)}{3(2)(x-1)} = \dfrac{4x-4-3x-3}{6(x+1)(x-1)} = \dfrac{x-7}{6(x-1)(x+1)}$

2. $\dfrac{(x+1)(x-2)}{(x+2)(x-2)} + \dfrac{(x-3)(x+2)}{(x-2)(x+2)} = \dfrac{x^2-x-2+x^2-x-6}{(x+2)(x-2)} = \dfrac{2(x^2-x-4)}{(x+2)(x-2)}$

3. $\dfrac{(x+1)(x-2)}{(x-1)(x+2)(x-2)} - \dfrac{(x+2)(x-1)}{(x-1)(x-2)(x+2)} = \dfrac{(x^2-x-2)-(x^2+x-2)}{(x-1)(x+2)(x-2)} = \dfrac{-2x}{(x-1)(x+2)(x-2)}$

4. $\dfrac{(2x+5)(2x-5)}{(x-4)(2x-5)} - \dfrac{(x+4)(x-4)}{(2x-5)(x-4)} = \dfrac{(4x^2-25)-(x^2-16)}{(x-4)(2x-5)} = \dfrac{3x^2-9}{(x-4)(2x-5)} = \dfrac{3(x^2-3)}{(x-4)(2x-5)}$

5. $\dfrac{(5x+2)(2x+1)}{(2x-1)(x-2)(2x+1)} + \dfrac{(x+1)(x-2)}{(2x-1)(2x+1)} = \dfrac{10x^2+9x+2+x^2-x-2}{(2x-1)(x-2)(2x+1)} = \dfrac{x(11x+8)}{(x-2)(2x-1)(2x+1)}$

6. $\dfrac{(x+2)(3x-4)}{(2x+3)(3x-4)} - \dfrac{(x-5)(2x+3)}{(3x-4)(2x+3)} + \dfrac{2x^2-10x-11}{(2x+3)(3x-4)} = \dfrac{3x^2+2x-8-(2x^2-7x-15)+2x^2-10x-11}{(2x+3)(3x-4)} =$

$\dfrac{3x^2-x-4}{(2x+3)(3x-4)} = \dfrac{(3x-4)(x+1)}{(2x+3)(3x-4)} = \dfrac{x+1}{2x+3}$

EXERCISE 6.4

1. $\dfrac{8}{3}$

$$1+\cfrac{1}{1-\cfrac{1}{1+\cfrac{1}{1-\cfrac{1}{1+\cfrac{1}{2}}}}} = 1+\cfrac{1}{1-\cfrac{1}{1+\cfrac{1}{1-\cfrac{1}{1+2}}}} = 1+\cfrac{1}{1-\cfrac{1}{1+\cfrac{1}{2}}} = 1+\cfrac{1}{1-\cfrac{1}{1+\cfrac{3}{2}}} = 1+\cfrac{1}{1-\cfrac{2}{5}} = 1+\cfrac{1}{\cfrac{3}{5}} = 1+\cfrac{5}{3} = \cfrac{8}{3}$$

2. $\dfrac{x+4}{x+2}$

$$\dfrac{\left(x+3-\dfrac{7}{x-3}\right)(x-3)}{\left(x+1-\dfrac{5}{x-3}\right)(x-3)} = \dfrac{x^2-9-7}{x^2-2x-3-5} = \dfrac{x^2-16}{x^2-2x-8} = \dfrac{(x+4)(x-4)}{(x+2)(x-4)} = \dfrac{x+4}{x+2}$$

3. $\dfrac{x+25}{3x+10}$

$$\dfrac{\left(5+\dfrac{123x}{x^2-10}\right)\left(x^2-10\right)}{\left(15-\dfrac{56x-170}{x^2-10}\right)\left(x^2-10\right)}=\dfrac{5x^2-50+123x}{15x^2-150-\left(56x-170\right)}=\dfrac{5x^2+123x-50}{15x^2-56x+20}=\dfrac{\left(5x-2\right)\left(x+25\right)}{\left(5x-2\right)\left(3x-10\right)}=$$

$$\dfrac{x+25}{3x-10}$$

4. $\dfrac{5}{7}$

$$\dfrac{\left(\dfrac{x+1}{x+6}-\dfrac{x-1}{x-6}\right)(x+6)(x-6)}{\left(\dfrac{x-1}{x+6}-\dfrac{x+1}{x-6}\right)(x+6)(x-6)}=\dfrac{(x+1)(x-6)-(x-1)(x+6)}{(x-1)(x-6)-(x+1)(x+6)}=\dfrac{x^2-5x-6-\left(x^2+5x-6\right)}{x^2-7x+6-\left(x^2+7x+6\right)}=$$

$$\dfrac{x^2-5x-6-x^2-5x+6}{x^2-7x+6-x^2-7x-6}=\dfrac{-10x}{-14x}=\dfrac{5}{7}$$

5. $\dfrac{2(x+2)}{x-13}$

$$\dfrac{\left(\dfrac{1}{5}+\dfrac{x+11}{x^2-49}\right)(10)\left(x^2-49\right)}{\left(\dfrac{1}{10}-\dfrac{x-1}{x^2-49}\right)(10)\left(x^2-49\right)}=\dfrac{2\left(x^2-49\right)+10(x+11)}{x^2-49-10(x-1)}=\dfrac{2x^2-98+10x+110}{x^2-49-10x+10}=$$

$$\dfrac{2x^2+10x+12}{x^2-10x-39}=\dfrac{2(x+3)(x+2)}{(x-13)(x+3)}=\dfrac{2(x+2)}{x-13}$$

EXERCISE 6.5

1. $x=-8,3$

$$\dfrac{4-x^2}{4-x}=-5$$

$$4-x^2=-5(4-x)$$

$$4-x^2=-20+5x$$

$$0=x^2+5x-24$$

$$(x+8)(x-3)=0$$

2. $x = 0, 10$

$$\left(\frac{x-5}{x-2} + \frac{2}{x-7} \right)(x-2)(x-7) = \left(\frac{31}{(x-2)(x-7)} \right)(x-2)(x-7)$$

$$(x-5)(x-7) + 2(x-2) = 31$$

$$x^2 - 12x + 35 + 2x - 4 = 31$$

$$x^2 - 10x = 0$$

$$x(x-10) = 0$$

3. $x = \frac{-13}{3}, 5$

$$x(x^2-1)\left(\frac{4x-2}{x^2-1} - \frac{1}{x} \right) = x(x^2-1)\left(\frac{66}{x(x^2-1)} \right)$$

$$x(4x-2) - (x^2-1) = 66$$

$$4x^2 - 2x - x^2 + 1 = 66$$

$$3x^2 - 2x - 65 = 0$$

$$(3x+13)(x-5) = 0$$

4. $x = -1, 6$

$$2x(x+2)\left(\frac{x+1}{2x} + \frac{x-4}{x+2} \right) = 2x(x+2)\left(\frac{5}{x} \right)$$

$$(x+2)(x+1) + 2x(x-4) = 10(x+2)$$

$$x^2 + 3x + 2 + 2x^2 - 8x = 10x + 20$$

$$3x^2 - 15x - 18 = 0$$

$$3(x^2 - 5x - 6) = 0$$

$$3(x-6)(x+1) = 0$$

5. $x = -1$

$$\frac{4}{(x-2)(x-6)} - \frac{x}{x-2} = \frac{1}{x-6}$$

$$(x-2)(x-6)\left(\frac{4}{(x-2)(x-6)} - \frac{x}{x-2} \right) = (x-2)(x-6)\left(\frac{1}{x-6} \right)$$

$$4 - x(x-6) = x-2$$

$$4 - x^2 + 6x = x - 2$$

$$0 = x^2 - 5x - 6$$

$$(x-6)(x+1) = 0$$

Reject $x = 6$

6. $x = -3, 5$

$$(3x-4)(x-3)\left(\frac{x+1}{3x-4}+\frac{2}{x-3}\right)=(3x-4)(x-3)\left(\frac{6x+4}{(3x-4)(x-3)}\right)$$

$$(x-3)(x+1)+2(3x-4)=6x+4$$

$$x^2-2x-3+6x-8=6x+4$$

$$x^2-2x-15=0$$

$$(x-5)(x+3)=0$$

7. $x = \dfrac{-7}{5}, 0, 4$

$$\frac{x+3}{x^2-1}=\frac{3x-5}{7x+2}+\frac{2x-1}{7x+2}$$

$$\frac{x+3}{x^2-1}=\frac{5x-6}{7x+2}$$

$$(x+3)(7x+2)=(x^2-1)(5x-6)$$

$$7x^2+23x+6=5x^3-6x^2-5x+6$$

$$5x^3-13x^2-28x=0$$

$$x(5x^2-13x-28)=0$$

$$x(5x+7)(x-4)=0$$

8. 9 hours

t: number of hours second pipe needs to fill the tank

$$\frac{4}{6}+\frac{3}{t}=1$$

$$4t+18=6t$$

9. 8.25 hours

t: time needed to complete the job

$$\frac{t+3}{25}+\frac{t}{15}=1$$

$$150\left(\frac{t+3}{25}+\frac{t}{15}\right)=1(150)$$

$$6(t+3)+10t=150$$

$$16t=132$$

10. 4 mph

r: rate of paddling in still water

$$\frac{10}{r+2}+\frac{10}{r-2}=\frac{20}{3}$$

$$30(r-2)+30(r+2)=20(r^2-4)$$

$$30r-60+30r+60=20r^2-80$$

$$20r^2-60r-80=0$$

$$20(r^2-3r-4)=0$$

$$20(r-4)(r+1)=0$$

EXERCISE 6.6

1. $4x^4y^6$

$$\left(2^7\right)^{\frac{2}{7}}\left(x^{14}\right)^{\frac{2}{7}}\left(y^{21}\right)^{\frac{2}{7}}=2^2x^4y^6$$

2. $\dfrac{27a^{12}}{8b^3}$

$$\left(\frac{3^4a^{16}}{2^4b^4}\right)^{\frac{3}{4}}=\frac{3^3a^{12}}{2^3b^3}$$

3. $\dfrac{25x^4z^5}{2}$

$$\frac{\left(5^3x^9z^8\right)^{\frac{2}{3}}}{\left(2^6x^{12}z^2\right)^{\frac{1}{6}}}=\frac{5^2x^6z^{\frac{16}{3}}}{2x^2z^{\frac{1}{3}}}=\frac{25x^4z^{\frac{15}{3}}}{2}$$

4. $80c^6$

$$\left(10^2b^4c^6\right)^{\frac{1}{2}}\left(8^3b^{-6}c^9\right)^{\frac{1}{3}}=\left(10b^2c^3\right)\left(8b^{-2}c^3\right)$$

5. $36m^6n^{12}$

$$\left(9^2m^{12}n^{24}\right)^{\frac{1}{2}}+\left(3^2m^4n^8\right)^{\frac{3}{2}}=9m^6n^{12}+27m^6n^{12}$$

EXERCISE 6.7

1. $10x^3 y^3 \sqrt[3]{y^2}$

$$\left(10^3 x^9 y^9 y^2\right)^{\frac{1}{3}} = 10x^3 y^3 \sqrt[3]{y^2}$$

2. $2x^2 z^5 \sqrt[4]{2xz}$

$$\left(2^4 x^8 z^{20}\right)^{\frac{1}{4}} \left(2xz\right)^{\frac{1}{4}} = \left(2x^2 z^5\right) \sqrt[4]{2xz}$$

3. $\dfrac{6x^2}{y}$

$$\left(3x^2 y^4\right)^{\frac{1}{3}} \left(72x^4 y^{-7}\right)^{\frac{1}{3}} = \left(216x^6 y^{-3}\right)^{\frac{1}{3}} = \left(6^3 x^6 y^{-3}\right)^{\frac{1}{3}} = \frac{6x^2}{y}$$

4. $2a^3 \sqrt[4]{2}$

$$\frac{\left(2^5 a^9 b^7\right)^{\frac{1}{2}}}{\left(2^5 a^6 b^{14}\right)^{\frac{1}{4}}} = \frac{2^{\frac{5}{2}} a^{\frac{9}{2}} b^{\frac{7}{2}}}{2^{\frac{5}{4}} a^{\frac{3}{2}} b^{\frac{7}{2}}} = 2^{\frac{5}{4}} a^3 = 2a^3 \sqrt[4]{2}$$

EXERCISE 6.8

1. 3

$$5x + 1 = 2^4$$

2. –3

$$21 - 5x = 6^2$$

3. 4

$$7x - 3 = 5^2$$

4. 10

$$\sqrt{3x+6} = x - 4$$
$$3x + 6 = (x-4)^2$$
$$3x + 6 = x^2 - 8x + 16$$
$$x^2 - 11x + 10 = 0$$
$$(x-1)(x-10) = 0$$
Reject $x = 1$

5. –7

$$\sqrt{11-2x} = x + 12$$
$$11 - 2x = (x+12)^2$$
$$11 - 2x = x^2 + 24x + 144$$
$$x^2 + 26x + 133 = 0$$
$$(x+7)(x+19) = 0$$
Reject $x = -19$

6. 4

$$\sqrt{x^2+9} = 2x - 3$$
$$x^2 + 9 = (2x-3)^2$$
$$x^2 + 9 = 4x^2 - 12x + 9$$
$$3x^2 - 12x = 0$$
$$3x(x-4) = 0$$
Reject $x = 0$

7. 3

These terms are a bit large. Use your graphing utility to find the solution.

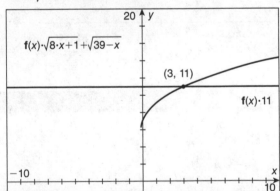

8. 9

$$\sqrt{6x+27}=5+\sqrt{x+7}$$

$$6x+27=\left(5+\sqrt{x+7}\right)^2$$

$$6x+27=25+10\sqrt{x+7}+x+7$$

$$10\sqrt{x+7}=5x-5$$

$$2\sqrt{x+7}=x-1$$

$$4(x+7)=(x-1)^2$$

$$4x+28=1-2x+x^2$$

$$x^2-6x-27=0$$

$$(x-9)(x+3)=0$$

Reject $x=-3$

9. 6

$$4x+3=3^3$$

10. 5

Use your graphing utility.

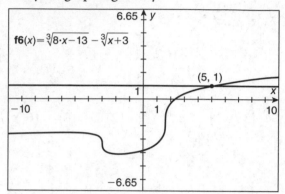

EXERCISE 6.9

1. $x<-2$ or $1<x<2$

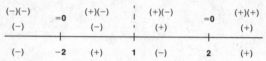

2. $x<-2$ or $-1<x<2$ or $x>5$

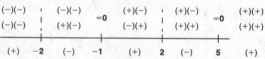

3. $x\le 0$ and $x\ne -2$ or $4<x\le 6$

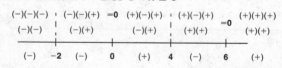

4. $x\le -4$ or $-3<x<3$ or $x\ge 4$

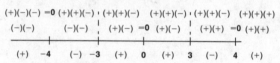

5. $x<\dfrac{-5}{3}$ or $0<x<1$ or $x>3$

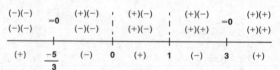

EXERCISE 6.10

1. 11.86 years

$$t^2=(1)(5.2)^3$$

2. 120 kg

$$125=50k \text{ tells us } k=2.5$$

$$300=2.5w$$

3. 54 inches

$$(162)(10)=30d$$

4. 20,000 dynes

$$\frac{(2)^2(5000)}{(40)(25)}=\frac{(1.5)^2(F)}{(30)(75)}$$

$$F=\frac{(2)^2(5000)(30)(75)}{(40)(25)(1.5)^2}$$

5. 675 cc

$$\frac{(1.5)(400)}{(2)(300)}=\frac{(2)(V)}{(3)(450)}$$

$$V=\frac{(1.5)(400)(3)(450)}{(2)(300)(2)}$$

CHAPTER 7

Exponential and Logarithmic Functions

EXERCISE 7.1

1. –1

$$(2)^{3-2x}=32=2^5$$
$$3-2x=5$$

2. 13

$$\left(3^2\right)^{5x+3}=\left(3^4\right)^{3x-5}$$
$$3^{10x+6}=3^{12x-20}$$
$$10x+6=12x-20$$
$$26=2x$$

3. 7

$$\left(2^3\right)^{4x-3}=\left(2^5\right)^{2x+1}$$
$$2^{12x-9}=2^{10x+5}$$
$$12x-9=10x+5$$
$$2x=14$$

4. 2.059

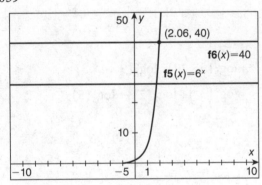

5. 5.19

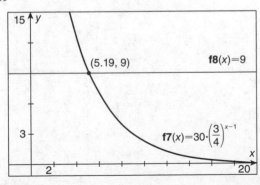

6. 20.4

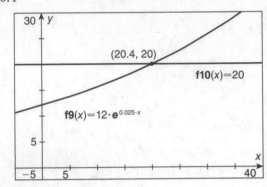

7. $15,089.58

$$A=10,000(1.042)^{10}$$

8. $15,186.33

$$A=10,000\left(1+\frac{.042}{4}\right)^{10\times4}=10,000(1.0105)^{40}$$

9. $15,208.46

$$A=10,000\left(1+\frac{.042}{12}\right)^{10\times12}=10,000\left(1+\frac{.042}{12}\right)^{120}$$

10. $1,847.06

$$V=3,500(1-0.12)^5=3,500(0.88)^5$$

EXERCISE 7.2

1. $4x + 2y - \dfrac{5}{2}z$

$$\log_b\left(m^4\right) + \log_b\left(n^2\right) - \log_b\left(\sqrt{p^5}\right)$$

$$4\log_b(m) + 2\log_b(n) - \frac{5}{2}\log_b(p)$$

$$4x + 2y - \frac{5}{2}z$$

2. $3x - 2y - \dfrac{2}{3}z$

$$\log_b\left(\frac{m^3}{n^2\sqrt[3]{P^2}}\right)$$

$$\log_b\left(m^3\right) - \log_b\left(n^2 p^{\frac{2}{3}}\right)$$

$$3\log_b(m) - \log_b\left(n^2\right) - \log_b\left(p^{\frac{2}{3}}\right)$$

$$3\log_b(m) - 2\log_b(n) - \frac{2}{3}\log_b(p)$$

$$3x - 2y - \frac{2}{3}z$$

3. $\dfrac{3}{2}x - \dfrac{3}{8}z$

$$\frac{1}{2}\log_b\left(\frac{m^3}{\sqrt[4]{p^3}}\right)$$

$$\frac{1}{2}\left(\log_b\left(m^3\right) - \log_b\left(p^{\frac{3}{4}}\right)\right)$$

$$\frac{1}{2}\left(3\log_b(m) - \frac{3}{4}\log_b(p)\right)$$

$$\frac{1}{2}\left(3x - \frac{3}{4}z\right) = \frac{3}{2}x - \frac{3}{8}z$$

4. $\dfrac{3x}{2z}$

$$\frac{\log_b\left(m^3\right)}{\log_b\left(p^2\right)}$$

$$\frac{3\log_b(m)}{2\log_b(p)} = \frac{3x}{2z}$$

5. $3x + y$

$$\log_b\left(8 \times 3\right)$$

$$3\log_b(2) + \log_b(3)$$

$$3x + y$$

6. $z - 2x - y$

$$\log_b\left(\frac{5}{4 \times 3}\right)$$

$$\log_b(5) - \log_b(4 \times 3)$$

$$\log_b(5) - \log_b\left(2^2\right) - \log_b(3)$$

$$z - 2x - y$$

7. $x + y + z + 2$

$$\log_b\left(2 \times 3 \times 5b^2\right)$$

$$\log_b(2) + \log_b(3) + \log_b(5) + \log_b\left(b^2\right)$$

$$x + y + z + 2\log_b(b)$$

$$x + y + z + 2(1) = x + y + z + 2$$

8. $2x + 2y + z$

$$\log_b(9 \times 20) = \log_b(9 \times 4 \times 5)$$

$$\log_b\left(3^2\right) + \log_b\left(2^2\right) + \log_b(5)$$

$$2y + 2x + z$$

9. $\dfrac{\log_b(217)}{\log_b(12)}$ or 2.16503

$$x\log_{12}(12)=\log_{12}(217)$$

$$x=\log_{12}(217)=\frac{\log(217)}{\log(12)}=\frac{\ln(217)}{\ln(12)}$$

10. 1.60593

$$7^{2x-1}=74$$

$$(2x-1)\log(7)=\log(74)$$

$$2x-1=\frac{\log(74)}{\log(7)}$$

$$x=\frac{\dfrac{\log(74)}{\log(7)}+1}{2}$$

11. $-3<x<-1$ or $x>4$

$$\frac{(x-4)(x+3)}{x+1}>0$$

$(-)\,(-)$	$=\mathbf{0}$	$(+)\,(-)$	$(+)\,(+)$	$=\mathbf{0}$	$(+)\,(+)$
$(-)$		$(-)$	$(-)$		$(+)$
$(-)$	$\mathbf{-3}$	$(+)$	$\mathbf{-1}$ $(-)$	$\mathbf{4}$	$(+)$

EXERCISE 7.3

1. 3.32

Use your calculator to compute $\dfrac{\log(210)}{\log(5)}$

2. 3.61

$$5^{3x-7}=468$$

$$(3x-7)\log(5)=\log(468)$$

$$3x-7=\frac{\log(468)}{\log(5)}$$

Use your calculator to compute $\dfrac{\log(468)}{\log(5)}$

Add 7 and then divide by 3.

3. 4.72

$$17(2.3)^x=870$$

$$(2.3)^x=\frac{870}{17}$$

$$x\log(2.3)=\log\left(\frac{870}{17}\right)$$

$$x=\frac{\log\left(\dfrac{870}{17}\right)}{\log(2.3)}$$

4. 6.35

$$19.3e^{-0.24x}=4.2$$

$$e^{-0.24x}=\frac{4.2}{19.3}$$

$$-0.24x=\ln\left(\frac{4.2}{19.3}\right)$$

$$x=\frac{\ln\left(\dfrac{4.2}{19.3}\right)}{-0.24}$$

5. -5, 13

$$x^2-8x-1=2^6=64$$

$$x^2-8x-65=0$$

$$(x-13)(x+5)=0$$

6. 10, 12

$$\frac{x^2+42x-8}{x-2}=4^3=64$$

$$x^2+42x-8=64(x-2)$$

$$x^2+42x-8=64x-128$$

$$x^2-22x+120=0$$

$$(x-12)(x-10)=0$$

7. –1.51, 1.77

Use your graphing utility.

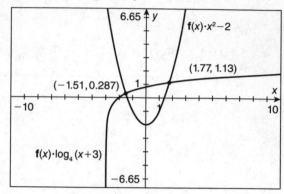

8. 9.17 mCi

$$A(24)=10e^{-.00361(24)}$$

9. 192.01 hours

$$5=10e^{-.00361t}$$

$$\frac{1}{2}=e^{-.00361t}$$

$$\ln\left(\frac{1}{2}\right)=-.00361t$$

$$t=\frac{\ln\left(\frac{1}{2}\right)}{-.00361}$$

CHAPTER 8

Sequences and Series

EXERCISE 8.1

1. 55

$1 + 2 + 3 + 4 + 5 + 6 + 7 + 8 + 9 + 10$

2. 385

$1 + 4 + 9 + 16 + 25 + 36 + 49 + 64 + 81 + 100$

3. 275

$-1 + 0 + 3 + 8 + 15 + 24 + 35 + 48 + 80$

4. 2,046

$2 + 4 + 8 + 16 + 32 + 64 + 128 + 256 + 512 + 1,024$

5. 2321

$$\sum_{n=1}^{10} \left(2^n + n^2 - 2n\right) = \sum_{n=1}^{10} 2^n + \sum_{n=1}^{10} n^2 - 2\sum_{n=1}^{10} n =$$

$2046 + 385 - 2(55)$

EXERCISE 8.2

1. 7, 10, 13, 16, 19

2. 63, 189, 567, 1,701, 5,103

3. 2,280, 2,160, 2,040, 1,920, 1,800

4. 200,000, 100,000, 50,000, 25,000, 12,500

5. 9, 13, 17, 21, 25, 29

$a_2 = a_1 + 4 = 9 + 4 = 13; a_3 = a_2 + 4 = 13 + 4 = 17$

6. 9, 31, 97, 295, 889, 2,671

7. 9, 15, 33, 87, 249, 735

8. 2, 7, 23, 76, 251, 829

$a_3 = 3a_2 + a_1 = 3(7) + 2 = 23$

9. f(n) = 7n − 4

There is a constant sum of 7 from term to term, therefore $f(n) = 7n + b$. Use the first term: $3 = 7(1) + b$ to find that $b = -4$.

10. $a_1 = 3$, $a_n = 6a_{n-1}$

There is a constant factor of 6 from term to term, so $a_1 = 3$ and $a_n = 6a_{n-1}$

EXERCISE 8.3

1. 57

$f(n) = 2n + 7; f(25) = 57$

2. 1,047

$f(n) = mn + b; f(19) = 243; f(11) = 147$

Solve the system of equations

$19m + b = 243$

$11m + b = 147$

$m = 12$ and $b = 15$

3. −4,930

The common difference is −163 so $f(n) = -163n + b$. Since $f(1) = 1,916$, $1,916 = -163 + b$ giving $b = 2,079$. $f(43) = -163(43) + 2,079$.

4. 1,334

The common difference is 36, so $f(n) = 36n + b$. With $f(1) = 146$, $b = 110$. $f(34) = 36(34) + 110$.

EXERCISE 8.4

1. 23,000

$$f(1) = 11; f(80) = 564; S_{80} = \frac{80}{2}(11+564)$$

2. 34,290

$$f(1) = 11; f(60) = 1,132; S_{60} = \frac{60}{2}(11+1,132)$$

3. 6,370

$$f(15) = 92; f(50) = 267; f(1) = 22;$$

$$S_{50} = \frac{50}{2}(22+267); S_{15} = \frac{15}{2}(22+92);$$

$$\sum_{n=16}^{50} 5n+17 = S_{50} - S_{15}$$

4. 825

$$S_{Left} = \frac{25}{2}(9+57)$$

5. 1125

$$S_{Middle} = \frac{25}{2}(21+69)$$

6. 875

Last row, right section: $f(n) = 2n + 9$; $f(25) = 59$

$$S_{Right} = \frac{25}{2}(11+59)$$

EXERCISE 8.5

1. 10,616.78

$$f(13) = 10,000(1.005)^{12}$$

2. 98,304

Solve the system of equations $48 = ar^3$ and $76 = ar^7$ by dividing: $\frac{768}{48} = 16 = \frac{ar^7}{ar^3} = r^4$. Therefore, $r = 2$ and $a = 6$. $f(15) = 6(2)^{14}$

3. 28

(1) Enter the function $f(n)=75(0.6)^{n-1}$ into your calculator (using an x instead of an n). Use the table of values to determine when $f < 0.0001$.

(2) Solve the equation $75(0.6)^{n-1} = 0.0001$ using logarithms. Since n must be an integer for this answer, round up on the decimal found (27.48) to get the correct number of terms.

4. 2.25×10^{14} mm or 2.25×10^{11} km

The thickness after n folds is given by $f(n) = .1(2)^{n+1}$. Therefore, $f(50) = .1(2)^{51}$.

(FYI: The average distance from the Earth to the Moon is 3.8×10^6 km.)

EXERCISE 8.6

1. 531,440

$$S_{12} = \frac{2(1-3^{12})}{1-3}$$

2. $\dfrac{531,440}{177,147}$

$$S_{12} = \frac{2\left(1-\left(\frac{1}{3}\right)^{12}\right)}{1-\frac{1}{3}}$$

3. $S_n = \dfrac{8\left(1-\left(\dfrac{3}{2}\right)^{10}\right)}{1-\dfrac{3}{2}} = \dfrac{58,025}{64}$

4. 3

$S_\infty = \dfrac{2}{1-\dfrac{1}{3}}$

5. $\dfrac{3}{2}$

$S_\infty = \dfrac{2}{1-\dfrac{-1}{3}}$

6. 198,426.66 (rounded to the nearest hundredth)

$S_{49} = \dfrac{1,000\left(1-1.05^{49}\right)}{1-1.05}$

7. The first deposit will grow interest for 49 years, the second deposit will grow interest for 48 years, and so on when she makes her last deposit on her birthday. This process is called an annuity.

CHAPTER 9

Trigonometry—Unit Circle and Triangles

EXERCISE 9.1

1. $\sin(\alpha)=\dfrac{4}{5}$; $\tan(\alpha)=\dfrac{4}{3}$; $\sec(\alpha)=\dfrac{5}{3}$; $\csc(\alpha)=\dfrac{5}{4}$; $\cot(\alpha)=\dfrac{3}{4}$

$\left(\dfrac{3}{5}\right)^2+\sin^2(\alpha)=1 \Rightarrow \sin(\alpha)=\dfrac{4}{5}$; $\sec(\alpha)=\dfrac{1}{\cos(\alpha)}=\dfrac{5}{3}$; $\csc(\alpha)=\dfrac{1}{\sin(\alpha)}=\dfrac{5}{4}$

$\tan(\alpha)=\dfrac{\sin(\alpha)}{\cos(\alpha)}=\dfrac{4}{3}$; $\cot(\alpha)=\dfrac{1}{\tan(\alpha)}=\dfrac{3}{4}$

2. $\sin(\beta)=\dfrac{5}{13}$; $\cos(\beta)=\dfrac{12}{13}$; $\tan(\beta)=\dfrac{5}{12}$; $\sec(\beta)=\dfrac{13}{12}$; $\cot(\beta)=\dfrac{12}{5}$

$\left(\dfrac{5}{13}\right)^2+\cos^2(\alpha)=1 \quad \cos(\alpha)=\dfrac{12}{13}$; $\sec(\alpha)=\dfrac{1}{\cos(\alpha)}=\dfrac{13}{12}$; $\csc(\alpha)=\dfrac{1}{\sin(\alpha)}=\dfrac{13}{5}$

$\tan(\alpha)=\dfrac{\sin(\alpha)}{\cos(\alpha)}=\dfrac{5}{12}$; $\cot(\alpha)=\dfrac{1}{\tan(\alpha)}=\dfrac{12}{5}$

3. $\sin(\theta)=\dfrac{7}{25}$; $\cos(\theta)=\dfrac{24}{25}$; $\sec(\theta)=\dfrac{25}{24}$; $\csc(\theta)=\dfrac{25}{7}$; $\cot(\theta)=\dfrac{24}{7}$

4. $\cos(\omega)=\dfrac{9}{41}$; $\tan(\omega)=\dfrac{40}{9}$; $\sec(\omega)=\dfrac{41}{9}$; $\csc(\omega)=\dfrac{41}{40}$; $\cot(\omega)=\dfrac{9}{40}$

5. $\cos(A)=\dfrac{2}{3}$; $\sin(A)=\dfrac{\sqrt{5}}{3}$; $\tan(A)=\dfrac{\sqrt{5}}{2}$; $\csc(A)=\dfrac{3}{\sqrt{5}}=\dfrac{3\sqrt{5}}{5}$; $\cot(A)=\dfrac{2}{\sqrt{5}}=\dfrac{2\sqrt{5}}{5}$

$\left(\dfrac{2}{3}\right)^2+\sin^2(\alpha)=1 \Rightarrow \sin^2(\alpha)=\dfrac{5}{9} \Rightarrow \sin(\alpha)=\dfrac{\sqrt{5}}{3}$

EXERCISE 9.2

1. $-\sin(46°)$

Quad IV: $314 = 360 - 46$

2. $-\tan(65°)$

Quad II: $-245 = -180 + (-65)$

3. $-\sec(88°)$

Quad III: $268 = 180 + 88$

4. $\csc(70°)$

Quad II: $110 = 180 - 70$

5. $-\cos(40°)$

Quad III: $-140 = -180 + 40$

6. $\dfrac{-1}{a}$

$$\cos(\theta)=a \Rightarrow \sec(\theta)=\frac{1}{a}; 180+\theta \to \text{QIII}; \sec(180+\theta)=\frac{-1}{a}$$

7. $\dfrac{-\sqrt{1-b^2}}{b}$

$$\sin(\theta)=b \Rightarrow \csc(\theta)=\frac{1}{b}; 360-\theta \to \text{QIV}; \csc(360-\theta)=\frac{-1}{b}; 1+\cot^2(360-\theta)=\csc^2(360-\theta)$$

$$\cot^2(360-\theta)=\frac{1}{b^2}-1=\frac{1-b^2}{b^2} \Rightarrow \cot(360-\theta)=\frac{-\sqrt{1-b^2}}{b}$$

8. $\dfrac{-1}{\sqrt{1+c^2}}=\dfrac{-\sqrt{1+c^2}}{1+c^2}$

$$\tan(\theta)=c \Rightarrow 1+c^2=\sec^2(\theta); 180-\theta \to \text{QII} \Rightarrow \sec(\theta)=-\sqrt{1+c^2} \Rightarrow \cos(180-\theta)=\frac{-1}{\sqrt{1+c^2}}$$

9. $\sqrt{1-\left(\dfrac{1}{d}\right)^2}=\dfrac{\sqrt{1-d^2}}{d}$

$$\csc(\theta)=d \Rightarrow \sin^2(\theta)=\frac{1}{d^2}; -\theta \to \text{QIV} \Rightarrow \cos^2(-\theta)+\frac{1}{d^2}=1 \Rightarrow \cos^2(-\theta)=1-\frac{1}{d^2}=\frac{1-d^2}{d^2}; \cos(-\theta)=\frac{\sqrt{1-d^2}}{d}$$

10. $\dfrac{-1}{\sqrt{e^2+1}}=\dfrac{-\sqrt{e^2+1}}{e^2+1}$

$$\cot(\theta)=e \Rightarrow 1+e^2=\csc^2(\theta) \Rightarrow \csc(\theta)=\sqrt{1+e^2}; -\theta \to \text{QIV} \Rightarrow \csc(-\theta)=-\sqrt{1+e^2} \Rightarrow \sin(-\theta)=\frac{-1}{\sqrt{1+e^2}}$$

		$\sin(\theta)$	$\cos(\theta)$	$\tan(\theta)$	$\csc(\theta)$	$\sec(\theta)$	$\cot(\theta)$
11.	0°	0	1	0	–	1	–
12.	30°	$\dfrac{1}{2}$	$\dfrac{\sqrt{3}}{2}$	$\dfrac{\sqrt{3}}{3}$	2	$\dfrac{2\sqrt{3}}{3}$	$\sqrt{3}$
13.	45°	$\dfrac{\sqrt{2}}{2}$	$\dfrac{\sqrt{2}}{2}$	1	$\sqrt{2}$	$\sqrt{2}$	1
14.	60°	$\dfrac{\sqrt{3}}{2}$	$\dfrac{1}{2}$	$\sqrt{3}$	$\dfrac{2\sqrt{3}}{3}$	2	$\dfrac{\sqrt{3}}{3}$
15.	90°	1	0	–	1	–	0
16.	120°	$\dfrac{\sqrt{3}}{2}$	$\dfrac{-1}{2}$	$-\sqrt{3}$	$\dfrac{2\sqrt{3}}{3}$	-2	$\dfrac{-\sqrt{3}}{3}$
17.	135°	$\dfrac{\sqrt{2}}{2}$	$\dfrac{-\sqrt{2}}{2}$	-1	$\sqrt{2}$	$-\sqrt{2}$	-1
18.	150°	$\dfrac{1}{2}$	$\dfrac{-\sqrt{3}}{2}$	$\dfrac{-\sqrt{3}}{3}$	2	$\dfrac{-2\sqrt{3}}{3}$	$-\sqrt{3}$
19.	180°	0	-1	0	–	-1	–
20.	210°	$\dfrac{-1}{2}$	$\dfrac{-\sqrt{3}}{2}$	$\dfrac{\sqrt{3}}{3}$	-2	$\dfrac{-2\sqrt{3}}{3}$	$\sqrt{3}$
21.	225°	$\dfrac{-\sqrt{2}}{2}$	$\dfrac{-\sqrt{2}}{2}$	1	$-\sqrt{2}$	$-\sqrt{2}$	1
22.	240°	$\dfrac{-\sqrt{3}}{2}$	$\dfrac{-1}{2}$	$\sqrt{3}$	$\dfrac{-2\sqrt{3}}{3}$	-2	$\dfrac{\sqrt{3}}{3}$
23.	270°	-1	0	–	-1	–	0
24.	300°	$\dfrac{-\sqrt{3}}{2}$	$\dfrac{1}{2}$	$-\sqrt{3}$	$\dfrac{-2\sqrt{3}}{3}$	2	$\dfrac{-\sqrt{3}}{3}$
25.	315°	$\dfrac{-\sqrt{2}}{2}$	$\dfrac{\sqrt{2}}{2}$	-1	$-\sqrt{2}$	$\sqrt{2}$	-1
26.	330°	$\dfrac{-1}{2}$	$\dfrac{\sqrt{3}}{2}$	$\dfrac{-\sqrt{3}}{3}$	-2	$\dfrac{2\sqrt{3}}{3}$	$-\sqrt{3}$
27.	360°	0	1	0	–	1	–

EXERCISE 9.3

1. $\dfrac{5\pi}{4}$

$$45° \leftrightarrow \dfrac{\pi}{4}^{r}$$

$$5(45° \pm) \leftrightarrow 5\left(\dfrac{\pi}{4}^{r}\right)$$

2. 4π

3. $\dfrac{-3\pi}{4}$

4. $\dfrac{4\pi}{5}$

$$\dfrac{x}{144} = \dfrac{\pi}{180} \Rightarrow x = \dfrac{144\pi}{180}$$

5. $240°$

6. $330°$

7. $315°$

8. $-120°$

9. $\dfrac{\sqrt{3}}{2}$

10. $\dfrac{-\sqrt{2}}{2}$

11. 1

12. $\dfrac{\sqrt{3}}{3}$

13. $\dfrac{2\sqrt{3}}{3}$

14. $\dfrac{2\sqrt{3}}{3}$

15. $\dfrac{-\sqrt{2}}{2}$

EXERCISE 9.4

1. $A = 8$, $p = \dfrac{\pi}{2}$, max = 13, min = –3

$$pd = \dfrac{2\pi}{|B|} = \dfrac{2\pi}{4} = \dfrac{\pi}{2}$$

$$\text{max} = 8 + 5$$

$$\text{min} = 8 - 5$$

2. $A = 5$, $p = \dfrac{1}{2}$, max = 8, min = –2

3. $A = 4$, $p = 4$, max = 3, min = –5

$$pd = \dfrac{2\pi}{\dfrac{\pi}{2}} = (2\pi)\left(\dfrac{2}{\pi}\right) = 4$$

4. $A = 3$, $p = 2$, max = 2, min = –4

$$pd = 2 - 0 = 2; amp = \dfrac{2 - (-4)}{2} = 3$$

5. $A = 5$, $p = 6\pi$, max = 7, min = –3

$\dfrac{3\pi}{2}$ is one-fourth of the period, so the period must equal $\left(\dfrac{3\pi}{2}\right)(4) = 6\pi$

6. $y = 3\cos(\pi x) - 1$

7. $y = -5\sin\left(\dfrac{1}{3}x\right) + 2$

The leading coefficient is –5 because the graph moves from its average at $x = 0$ to its minimum value.

EXERCISE 9.5

1. $\dfrac{-\pi}{3}$

$\tan(\theta) = -\sqrt{3}$ and θ in Quadrant II

2. $\dfrac{-\pi}{4}$

3. $\dfrac{2\pi}{3}$

$\sec^{-1}(-2) = \cos^{-1}\left(\dfrac{-1}{2}\right)$

4. $\dfrac{15}{17}$

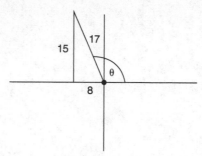

$\theta = \cos^{-1}\left(\dfrac{-8}{17}\right)$

5. $\dfrac{-\sqrt{7}}{3}$

EXERCISE 9.6

1. 126.9°, 233.1°

$\cos(\alpha) = \dfrac{-3}{5}$

Reference angle: $\cos^{-1}\left(\dfrac{3}{5}\right) = 53.1°$

2. 63.4°, 135°, 243.4°, 315°

$\left(\tan(\beta)+1\right)\left(\tan(\beta)-2\right) = 0$

$\tan(\beta) = -1, 2$

Reference angles:

$\tan(\beta) = -1 \rightarrow \beta = 45°$

$\tan(\beta) = 2 \rightarrow \beta = 63.4°$

3. 41.8°, 138.2°, 221.8°, 318.2°

$\sin(\omega) = \pm\dfrac{2}{3}$

Reference angle:

$\sin^{-1}\left(\dfrac{2}{3}\right) = 41.8°$

4. 41.4°, 131.8°, 228.2, 318.6°

$\left(4\cos(A)-3\right)\left(3\cos(A)+2\right) = 0$

$\cos(A) = \dfrac{3}{4}, \dfrac{-2}{3}$

Reference angles:

$\cos(A) = \dfrac{3}{4} \rightarrow A = 41.4°$

$\cos(A) = \dfrac{2}{3} \rightarrow A = 48.2°$

5. 80.3°, 279.7°

$3\sec^2(Q) - 17\sec(Q) - 5 = 0$

$\sec(Q) = \dfrac{17 \pm \sqrt{17^2 - 4(3)(-5)}}{6}$

$\sec(Q) = \dfrac{17 \pm \sqrt{349}}{6}$

$\sec(Q) = 5.947, -0.280$

Reject -0.280 because this is not in the range of the secant function.

Reference angle:

$\sec^{-1}(5.947) = \cos^{-1}\left(\dfrac{1}{5.947}\right) = 80.°$

6. 3.99, 5.44

$$\sin(\alpha) = \frac{-3}{4}$$

Reference angle:

$$\sin^{-1}\left(\frac{3}{4}\right) = 0.85$$

QIII: $\alpha = \pi + 0.85$

QIV: $\alpha = 2\pi - 0.85$

7. 0.59, 2.55, 3.73, 5.70

$$\tan(\beta) = \frac{-2}{3}$$

Reference angle:

$$\tan^{-1}\left(\frac{2}{3}\right) = 0.59$$

QII: $\beta = \pi - 0.59$

QIII: $\beta = \pi + 0.59$

QIV: $\beta = 2\pi - 0.59$

8. $1.57.\left(\frac{\pi}{2}\right)$, 3.79, 5.64

$$\left(5\sin(R) + 3\right)\left(\sin(R) - 1\right) = 0$$

$$\sin(R) = \frac{-3}{5}, 1$$

Reference angle:

$$\sin^{-1}\left(\frac{3}{5}\right) = 0.64$$

9. $0.25, 1.57\left(\frac{\pi}{2}\right), 2.89$

$$\left(\csc(\beta) - 1\right)\left(\csc(\beta) - 4\right) = 0$$

$$\csc(\beta) = 1, 4$$

$$\sin(\beta) = 1, \frac{1}{4}$$

Reference angle:

$$\sin^{-1}\left(\frac{1}{4}\right) = 0.25$$

10. 0.93, 2.68, 4.07, 5.82

$$\tan(\theta) = \frac{5 \pm \sqrt{5^2 - 4(6)(-4)}}{12}$$

$$\tan(\theta) = \frac{5 \pm \sqrt{121}}{12} = \frac{5 \pm 11}{12} = \frac{4}{3}, \frac{-1}{2}$$

Reference angles:

$$\tan^{-1}\left(\frac{4}{3}\right) = 0.93$$

$$\tan^{-1}\left(\frac{1}{2}\right) = 0.46$$

QII: $\beta = \pi - 0.46$

QIII: $\beta = \pi + 0.93$

QIV: $\beta = 2\pi - 0.46$

EXERCISE 9.7

1. $h(t) = 260\cos\left(\dfrac{\pi}{900}t\right) + 290$

Radius of the Ferris wheel is the amplitude of the sine wave: 260

Maximum height – radius = distance from ground to center of the wheel: 290

Period = 30 minutes = 1,800 seconds;

$B = B = \dfrac{2\pi}{1,800} = \dfrac{\pi}{900}$

2. $d(t) = -21\cos\left(\dfrac{\pi}{745}t\right) + 32$

Amplitude $= \dfrac{max - min}{2} = \dfrac{53 - 11}{2} = \dfrac{42}{2} = 21$

Average = max – amplitude = 32

Period = 24 hours 50 minutes = 1,490 minutes;

$B = \dfrac{2\pi}{1,490} = \dfrac{\pi}{745}$

This cosine graph has been reflected over its average because we begin the clock at the low tide.

3. $T(d) = -8\cos\left(\dfrac{2\pi}{365}d\right) + 76$

Amp $= \dfrac{max - min}{2} = \dfrac{84 - 68}{2} = \dfrac{16}{2} = 8$

Average = max – amp = 84 – 8 = 76

Period = 365 days

4. $d(t) = 120\sin\left(\dfrac{2\pi}{5}t\right) + 130$

CHAPTER 10

Descriptive Statistics

EXERCISE 10.1

1. Median = 28; mean = 30.9

The median is the fifth piece of data after the data has been arranged; the sum of the 9 pieces of data is 278.

2. Median = 31; mean = 39.8

The median is the mean of the fifth and sixth pieces of data after the data has been arranged; the sum of the 10 pieces of data is 398.

3. Median = $54 million; mean = $63.2 million

The median is the thirteenth piece of data after the data has been arranged; the sum of the 25 pieces of data is 1,580.5.

4. Median = $4 million; mean = $7.31 million

The median is the 218th piece of data after the data has been arranged; the sum of the 435 pieces of data is 3,179.

5. 23

$$\text{Solve } \frac{14(20)+16(30)+21(40)+n(50)+12(60)+14(70)}{77+n} = 44.5$$

EXERCISE 10.2

1. Range = 49; IQR = 24; s = 15.6

max = 61; min = 12, Q1 = 18, Q3 = 42

2. Range = 108; IQR = 26; s = 31.8

max = 120, min = 12, Q1 = 19, Q3 = 45

3. Range = $92 million; IQR = $41.75 million; s = $24.8 million

max = 130, min = 38, Q1 = 42, Q3 = 83.75

4. Range = $33 million; IQR = $9 million; σ = $7.42 million

max = 34, min = 1, Q1 = 2, Q3 = 11

5. Median = 600; Mean = 634.3; Range = 1090; IQR = 400; s = 275.0

The median is the mean of the twenty-seventh and twenty-eighth pieces of data; the sum of the data is 34,250; max = 1,310, min = 220, Q1 = 400, Q3 = 800

EXERCISE 10.3

1.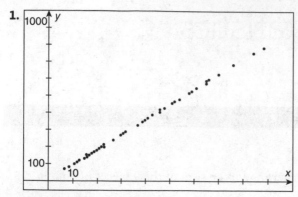

2. $fc = 9.01$ (fat) – 0.42

3. For each increase of 1 gram of fat in a sandwich the number of calories due to fat increases by 9.01.

4. 179.78 calories

5.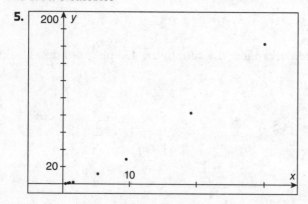

6. Power relationship

7. yr = 0.99999(dist)$^{1.50}$

8. 4.44yrs

9.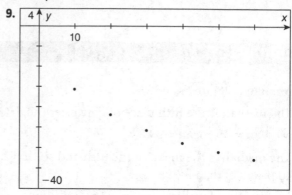

10. logarithmic

11. windchill = 6.46 – 9.59 ln(x)

12. –30.05°F

EXERCISE 10.4

1. 0.2995

  normalcdf(61,63.5,64,2.6)

2. 0.7781

  normalcdf(56,66.5,64,2.6)

3. 0.5762

  normalcdf(63.5,64,64,2.6) + 0.5

4. 0.7791

  normalcdf(0,66,64,2.6)

5. 0.2219

  1 – normalcdf(56,66.5,64,2.6)

6. 0.0444

  1 – normalcdf(58.5,69,64,2.6)

7. 0.7118

  normalcdf(15.78,16.2,15.87,0.15)

8. 0.9374

  normalcdf(0,16.1,15.87,0.15)

9. 0.0002

  .5 – normalcdf(15.87,16.4,15.87,0.15)

10. 0.4609

  1 – normalcdf(015.85,16.2,15.87,0.15)

11. 0.0359

  0.5 – normalcdf(4.3,4.75,4.3,0.25)

CHAPTER 11

Inferential Statistics

EXERCISE 11.1

1. No—the northeast is not necessarily representative of the entire country.

2. (1) Randomly select 1,000 people from around the country.

(2) Randomly select 100 people from each of the 50 states.

EXERCISE 11.2

1. 1.071

$$\frac{s}{\sqrt{n}} = \frac{7.5}{\sqrt{49}}$$

2. 2.164

$$\frac{s}{\sqrt{n}} = \frac{15.3}{\sqrt{50}}$$

3. 0.037

$$\sqrt{\frac{p(1-p)}{n}} = \sqrt{\frac{0.7(0.3)}{150}}$$

4. 0.022

$$\sqrt{\frac{p(1-p)}{n}} = \sqrt{\frac{0.5(0.5)}{500}}$$

5. 0.4058

$$\text{normalcdf}\left(68.5, 69.3, 69.3, \frac{4.3}{\sqrt{50}}\right)$$

6. 0.0942

$$0.5 - \text{normalcdf}\left(68, 69.3, 69.3, \frac{4.3}{\sqrt{50}}\right)$$

7. 0.1248

$$0.5 - \text{normalcdf}\left(69.3, 70, 69.3, \frac{4.3}{\sqrt{50}}\right)$$

8. 0.6851

$$\text{normalcdf}\left(0.985, 0.995, 0.99, \sqrt{\frac{.99(.01)}{400}}\right)$$

9. 0.1574

$$0.5 - \text{normalcdf}\left(0.99, 0.995, 0.99, \sqrt{\frac{.99(.01)}{400}}\right)$$

10. 0.1574

If less than 0.5% fail, then more than 99.5% meet the specifications. This is the same question as number 9.

EXERCISE 11.3

1. 2.5

$$\frac{35-25}{4}$$

2. 1.75

$$\frac{35-28}{4}$$

3. 1.6

$$\frac{35-31}{2.5}$$

4. $A = 69.8699$; $B = 86.5301$

$A = \text{invNorm}(.10, 78.2, 6.5)$

5. $A = 877.804$; $B = 1,002.2$

$A = \text{invNorm}(.20, 940, 73.9)$

6. $0.996737 \, L$

$A = \text{invNorm}(0.01, 1.02, 0.01)$

7. 15.9258 hands

$A = \text{invNorm}(0.96, 14, 1.1)$

EXERCISE 11.4

1. The personnel department can be 95% certain that the mean amount of money spent per employee for medical coverage is between $727.15 and $777.21.

zInterval 90.3,752.18,50,0.95: *stat.results*

"Title"	"z Interval"
"CLower"	727.151
"CUpper"	777.209
"$\bar{x}$"	752.18
"ME"	25.0294
"n"	50.
"σ"	90.3

2. The manager of the rock and roll band can be 90% certain that the average attendance is between 10,293 and 11,007 people per performance.

zInterval 1190,10650,30,0.9: *stat.results*

"Title"	"z Interval"
"CLower"	10292.6
"CUpper"	11007.4
"$\bar{x}$"	10650.
"ME"	357.366
"n"	30.
"σ"	1190.

3. The automobile dealership manager can be 98% certain that the proportion of leases for new transactions is between 42.0605% and 67.9395%.

zInterval_1Prop 44,80,0.98: *stat.results*

"Title"	"1 – Prop z Interval"
"CLower"	0.420605
"CUpper"	0.679395
"$\hat{p}$"	0.55
"ME"	0.129395
"n"	80.

EXERCISE 11.5

1. $H_0 : \mu \geq 5.02$

$H_a : \mu < 5.02$

CP: 5.01; p-value $= 7.07 \times 10^{-13}$

Reject H_o and claim that the true mean weight of the flour in a 5 lb. bag of flour is less than 5.02 lbs.

zTest 5.02,0.03,4.99,50,−1: *stat. results*	
"Title"	"z Test"
"Alternate Hyp"	"$\mu < \mu 0$"
"z"	−7.07107
"PVal"	7.73E-13
"$\bar{x}$"	4.99
"n"	50.
"σ"	0.03

2. $H_0 : p \geq 0.6$

$H_a : p < 0.6$

CP: .556; p-value $= 8.134 \times 10^{-9}$

Reject H_0 and claim that less than 60% of dentists recommend chewing sugarless gum.

zTest_1Prop 06,2225,4000,−1: *stat. results*	
"Title"	"1—Prop z Test"
"Alternate Hyp"	"prop $< p0$"
"z"	−5.6481
"PVal"	8.134E-9
"$\hat{p}$"	0.55625
"n"	4000.

3. $H_0 : \mu \leq 30.7$

$H_a : \mu > 30.7$

CP: 32.2501; p-value $= 0.2375$

Fail to reject H_0 and claim that not enough evidence has been found to reject the claim that there has been a decrease in the number of hours of overtime.

zTest 30.7,2.8,31.1,25,1: *stat. results*	
"Title"	"z Test"
"Alternate Hyp"	"$\mu > \mu 0$"
"z"	0.714286
"PVal"	0.237525
"$\bar{x}$"	31.1
"n"	25.
"σ"	2.8

4. $H_0 : p \geq 0.5$

$H_a : p < 0.5$

CP: 0.2674 ; p-value $= 0.3645$

Fail to reject H_0 and claim that not enough evidence has been found to reject the claim that more than half of the songs played on the radio and by Internet services are love songs.

zTest_1Prop 0.5,36,75,−1: *stat. results*	
"Title"	"1−Prop z Test"
"Alternate Hyp"	"prop $< p0$"
"z"	−0.34641
"PVal"	0.364517
"$\hat{p}$"	0.48
"n"	75.

EXERCISE 11.6

1. Identify the eight dinosaurs by the numbers 1–8. Randomly select an integer 1 through 8. The first selection will be a success, as you currently have none of the dinosaurs. For the second trial forward, continue picking a random integer in the interval 1–8 until a new dinosaur is picked. Continue the trials until you have all 8 dinosaurs. Repeat this experiment a number of times and find the average of these results.

2. Define the integers 0–5 as the Western Conference team winning a game and 6–9 as the Eastern Conference team winning. Randomly select 7 integers from 0–9. Identify each entry as a win for the appropriate conference until one of the teams is the first to get 4 wins. Repeat this process a number of times and compute the average number of games needed to determine the winner of the series.

An Introduction to Matrices

EXERCISE A.1

1. (−10, 15)

$$\begin{bmatrix} 6 & 4 \\ 4 & 9 \end{bmatrix} \begin{bmatrix} x \\ y \end{bmatrix} = \begin{bmatrix} 0 \\ 95 \end{bmatrix}$$

$$\begin{bmatrix} x \\ y \end{bmatrix} = \begin{bmatrix} 6 & 4 \\ 4 & 9 \end{bmatrix}^{-1} \begin{bmatrix} 0 \\ 95 \end{bmatrix} = \begin{bmatrix} -10 \\ 15 \end{bmatrix}$$

2. (8, 9, 14)

$$\begin{bmatrix} 2 & 3 & -2 \\ 3 & -4 & 2 \\ 7 & 10 & 5 \end{bmatrix} \begin{bmatrix} x \\ y \\ z \end{bmatrix} = \begin{bmatrix} 15 \\ 16 \\ 216 \end{bmatrix}$$

$$\begin{bmatrix} x \\ y \\ z \end{bmatrix} = \begin{bmatrix} 2 & 3 & -2 \\ 3 & -4 & 2 \\ 7 & 10 & 5 \end{bmatrix}^{-1} \begin{bmatrix} 15 \\ 16 \\ 216 \end{bmatrix} = \begin{bmatrix} 8 \\ 9 \\ 14 \end{bmatrix}$$

APPENDIX B

Conditional and Binomial Probabilities

1. $P(\text{Honor}|\text{Orchestra}) = \dfrac{21}{47}$

There are 21 students in the orchestra and the honor society. There are 47 students in orchestra.

2. $P(\overline{\text{Orchestra}}|\text{Honor}) = \dfrac{14}{35}$

Of the 35 students in the honor society, 14 are not members of the orchestra.

3. $P(\text{Honor} \cap \text{Orchestra}) = \dfrac{21}{143} = 0.147;$

$P(\text{Honor}) = \dfrac{35}{143} = 0.245;$

not independent

There are 143 (26 + 21 + 14+ 82) members in the senior class, 35 of whom are in the honor society. Because the probabilities of the two events are not equal, the events are not independent.

4. $P(1-2\,|\,15-17) = \dfrac{20}{72}$

Of the 72 people in the age group 15–17, 20 listen to music for 1–2 hours each day.

5. $P(18-20\,|\,4^{+}) = \dfrac{21}{67}$

Of the 67 people who listen to music for at least 4 hours each day, 21 are in the age range 18–20.

6. $P(21-24) = \dfrac{78}{222} = 0.351;$

$P(21-24\,|\,2-3) = \dfrac{34}{79} = 0.430;$

not independent

Of the 222 people in the survey, 78 are between the ages of 21 and 24. The probability that a person selected comes from this group does not equal the conditional probability that a person is drawn from this age group given that the person listens to 2 to 3 hours of music each day.

1. $P(r = 3) = \text{binompdf}(10,3,.5) = 0.1172$

2. $P(r \le 2) = \text{binomcdf}(5,\dfrac{1}{6},0,2) = 0.9645$

3. $P(r \ge 2) = \text{binomcdf}(8,\dfrac{1}{3},2,8) = 0.8049$

4. $P(r = 100) = \text{binompdf}(100,0.99,100) = 0.3660$

5. $P(r \le 3) = \text{binomcdf}(50,.01,0,3) = 0.9984$

6. $P(r \ge 99) = 1 - \text{binomcdf}(50,.01,0,98) = 0$